China's Information
Revolution

China's Information Revolution

Managing the Economic and Social Transformation

Christine Zhen-Wei Qiang

THE WORLD BANK
Washington, D.C.

© 2007 The International Bank for Reconstruction and Development / The World Bank
1818 H Street NW
Washington, DC 20433
Telephone: 202-473-1000
Internet: www.worldbank.org
E-mail: feedback@worldbank.org

All rights reserved

1 2 3 4 10 09 08 07

This volume is a product of the staff of the International Bank for Reconstruction and Development / The World Bank. The findings, interpretations, and conclusions expressed in this volume do not necessarily reflect the views of the Executive Directors of The World Bank or the governments they represent. The World Bank does not guarantee the accuracy of the data included in this work. The boundaries, colors, denominations, and other information shown on any map in this work do not imply any judgement on the part of The World Bank concerning the legal status of any territory or the endorsement or acceptance of such boundaries.

Rights and Permissions

The material in this publication is copyrighted. Copying and/or transmitting portions or all of this work without permission may be a violation of applicable law. The International Bank for Reconstruction and Development / The World Bank encourages dissemination of its work and will normally grant permission to reproduce portions of the work promptly.

For permission to photocopy or reprint any part of this work, please send a request with complete information to the Copyright Clearance Center Inc., 222 Rosewood Drive, Danvers, MA 01923, USA; telephone: 978-750-8400; fax: 978-750-4470; Internet: www.copyright.com.

All other queries on rights and licenses, including subsidiary rights, should be addressed to the Office of the Publisher, The World Bank, 1818 H Street NW, Washington, DC 20433, USA; fax: 202-522-2422; e-mail: pubrights@worldbank.org.

DOI: 10.1596/978-0-8213
Cover design: Quantum Think, Philadelphia, Pennsylvania

Library of Congress Cataloging-in-Publication Data

Qiang, Christine Zhen-Wei.
   China's information revolution: managing the economic and social transformation / by Christine Zhen-Wei Qiang.
       p. cm.
   Includes bibliographical references and index.
   ISBN-13: 978-0-8213-6720-9
   ISBN-10: 0-8213-6720-X
   ISBN-10: 0-8213-6721-8 (electronic)
   1. Information technology—China—Management. 2. China—Economic conditions—21st century. 3. China—Social conditions—21st century. 4. Social change—China—21st century. I. Title.
   HC430.T4Q53 2007
   338.4'70040951—dc22
                                                          2006103069

Contents

Foreword ix
Foreword xi
Preface xiii
Acknowledgments xv
Abbreviations xvii

Overview 1

1 China's Emerging Informatization Strategy 11

2 Establishing an Enabling Legal and Regulatory Environment 23

3 Enhancing Telecommunications Infrastructure 39

4 Developing and Innovating the ICT Industry 55

5 Improving ICT Human Resources 75

6 Advancing E-Government 89

7 Fostering E-Business 105

8 Connecting the Issues: A Summing Up 119

References 125

Boxes
1.1 Characteristics of Information and Communication Technology 12
2.1 China's E-Signature Law 29
2.2 The United Kingdom's Converged Regulator 33
2.3 Legislative Mechanisms in China 36
4.1 Government Initiatives toward the Integrated Circuit Industry 56
4.2 The Key Factor for Developing Integrated Circuit Design Capacity 57

4.3 Lenovo's Purchase of IBM   60
4.4 Leading Domestic Security Firms   65
4.5 TRLabs—An Industry-Led ICT R&D Consortium   71
5.1 Information Retrieval Abilities among Primary and Secondary School Students   77
5.2 IBM's Software Engineer Training and Certification Program   79
5.3 Launch of the Union of National Teachers Education Network   80
5.4 Training for Government CIOs   82
5.5 Beijing Raises Public Awareness of Informatization   86
6.1 A Government Web Site Is Not a One-Off Investment   96
6.2 Rural Informatization Case Study of Chongqing   99
7.1 Examples of Enterprises' Internal Informatization Applications   108
7.2 Large Firms Lead in B2B E-Commerce   112
7.3 Alibaba China   113

Figures

1.1 General Pattern of Informatization Strategy Development   14
1.2 China's Economic Structure by Sector, 1978–2003   17
1.3 Foreign Direct Investment in China, 1996–2005   18
1.4 China's Urban-Rural Population, 1995–2004, and Income Ratio, 1978–2004   19
1.5 Framework for China's Informatization   21
2.1 Institutional Structure of China's Telecommunications Sector   24
2.2 Regulatory Jurisdictions for China's ICT Services and Networks   26
3.1 Telecommunications Investment and Revenue in China, 1990–2004   40
3.2 Employees and Revenues of China's Main Telecommunications Providers   42
3.3 Market Shares of Fixed and Mobile Providers, 2005   43
3.4 Telecommunications Penetration in China, 1994–2004   44
3.5 Charges for Broadband (ADSL) Access in Beijing, 2001–03   45
3.6 Monthly Price Basket for Internet Use in Selected Countries, 2003   45
3.7 Fixed Line and Mobile Telephone Penetration by Region, 2003   46
3.8 Fixed Line and Internet Penetration in China's Urban and Rural Areas, 2003   47
3.9 Personal Computer Penetration in Selected Provinces in China, 2003   50
4.1 Sales Revenues for China's Integrated Circuit Industry by Segment, 2004   57
4.2 Top Global Producers of Computer Hardware, 1995 and 2000–04   59
4.3 Market Shares of the Top Six Personal Computer Firms in China, 2004   59
4.4 China's Software Market, 1999–2004   62
4.5 Network Security Revenue in China's Vertical Markets, Q1 2003–Q2 2004   64
4.6 Size and Growth of China's Digital Media Industry, 2001–05   66
4.7 R&D Spending and ICT Patent Applications in Selected Countries, 2004   72
5.1 Gross Secondary and Tertiary Enrollment Ratios in Selected Countries, 1980 and 2003   76

Contents

5.2 Regional Differences in ICT Education in China's Primary and Secondary Schools, Selected Provinces   77
5.3 Chinese People's Reasons for Not Using the Internet, 2005   81
6.1 Three Stages of E-Government Development in China   90
6.2 ICT Application Use by Chinese Government Departments, 2004   91
6.3 Implementation of Selected Golden Projects, 2004   94
6.4 Government Web Site Quality, 2005   96
6.5 E-Community Content in China, 2004   97
7.1 Informatization at Enterprises   106
7.2 Objectives of Informatization for Chinese Enterprises, 2003   106
7.3 Chinese Firms' Investments in Informatization by Industry, 2003   107
7.4 Internet Access and ICT Application Use in China's Manufacturing Industry, 2003   109
7.5 Prevalence and Reported Impact of ICT Applications in Chinese Firms, 2003   110
7.6 Changes in Supplier and Client Contacts among Chinese Firms Engaged in E-Commerce, 2003   111
7.7 Main Obstacles to E-Commerce in China, 2003   113
7.8 Frequency and Spending of Online Shoppers in China, 2005   115
7.9 Main Drawbacks to Online Purchases in China, 2005   115

Tables

2.1 ICT-Related Regulatory Responsibilities of Selected Government Agencies   25
2.2 Main Areas, Goals, and Policy, Legal, and Regulatory Issues for Informatization   27
2.3 E-Commerce and E-Signature Legislation in East Asia and the Pacific   29
2.4 Selected ICT Laws and Regulations in China   31
3.1 China's Main Telecommunications Providers, by Market Segment   41
3.2 China's Telecommunications Commitments to the World Trade Organization   42
3.3 Telecommunications Penetration in East Asia and Other Developing Economies, 2004   46
4.1 Chinese Software Parks   61
5.1 Annual Supply of and Demand for ICT Professionals in China, by Field   83
6.1 E-Government Readiness Rankings in East and South Asia, 2004 and 2005   92
6.2 China's Golden Projects   93
6.3 Top 10 Government Web Sites by Type of Sponsor, 2005   95

Foreword

Informatization—defined as the transformation of an economy and society driven by information and communication technology (ICT)—is not an end in itself but a complex process for achieving more critical development goals. This process involves investing significantly in economic and social infrastructure that facilitates the use of ICT by government, industry, civil society, and the general public. The long-term goal of informatization is to build an information society.

Since the 1980s ICT has increasingly been used to achieve economic and social goals. A variety of countries—both developed and developing—have made considerable progress in promoting informatization and fostering enabling environments for new technology.

Over the past decade China has also devoted considerable resources to informatization. Indeed, informatization and economic development have been mutually reinforcing. However, informatization efforts require updating to address the challenges and opportunities created by industrialization, urbanization, upgraded consumption, and increased social mobility. Developing a new, more effective informatization strategy will help China achieve its economic and social goals by spurring innovation, supporting more efficient use of economic resources, and increasing productivity and international competitiveness.

This publication is the result of 18 months of strategic research by a World Bank team, at the request of China's State Council Informatization Office and the Advisory Committee for State Informatization. Drawing on a half-dozen background papers by Chinese researchers, the study provides a variety of domestic perspectives and local case studies. By combining these perspectives with international experiences on how similar issues may have been addressed by other World Bank client countries, the report provides guidance on the kind of policies and reforms the authorities may wish to consider in pursuing China's quest for continued informatization.

Jim Adams
Vice President of the East Asia and Pacific Region
The World Bank

Foreword

As per request of the State Council Informatization Office and the Advisory Committee of the State Informatization, the World Bank Group, in cooperation with a number of Chinese experts, has successfully conducted research on the development strategy of China's informatization. The final report, *China's Information Revolution: Managing the Economic and Social Transformation,* summarizes the achievements of this research in a condensed form and presents a series of pertinent recommendations. In addition, many observations, findings, and suggestions of the draft report have played an important role in developing China's informatization strategies, which was conducted in 2005.

The Chinese government published the *National Informatization Strategy: 2006–2020* in May 2006 and reconfirmed that informatization is an integrated portion of China's national strategies for moving toward modernization. This strategic publication also clearly defines directive principles, strategic objectives, guiding policies, and primary action plans. With the strategy's implementation, there is no doubt that China's informatization will enter into a new phase and make even more significant contributions to China's economic and social development as well as the course of modernization.

The World Bank has had a long-term and effective collaboration with the Chinese government and has contributed substantially to assist China in achieving its goal of modernization. I sincerely hope that this cooperation will continue, not only for the benefit of Chinese people but also for the benefit of people in the developing world. I would like to take this opportunity to express our sincere thanks to the World Bank Group for its highly effective assistance, in particular, to Christine Zhen-Wei Qiang and her team for their creative endeavors, professionalism, and dedication.

Zhou Hongren, Ph.D.
Executive Vice Chairman
The Advisory Committee for State Informatization

Preface

A growing number of countries has recognized the high potential of information and communication technologies (ICT) to contribute to national economic and social development. As China's development has entered a new stage, it also requires an updated "informatization" strategy for its economic and social transformation.

China's new ICT strategy needs to meet the unprecedented opportunities—and challenges—posed by a fast-growing economy with large, diverse, and widely spread population. ICTs could effectively support institutional changes to make government functions more service-oriented, efficient, and transparent. In doing so, it would make markets and resource allocations far more efficient to sustain growth. ICTs could also help reform manufacturing and energy industries, reducing the cost of capital and increasing the value added of Chinese products, as well as enhancing Chinese enterprises' productivity, international competitiveness, and capacity for technological innovation in a broad range of products and processes. The challenge is how to ensure that the deployment of ICTs would contribute to reducing disparity and bringing about a more balanced and equitable social and economic development to all regions of the country.

Like many countries, China faces the challenge of adapting its policies to fast-moving technologies and institutional models. In such a context, policies should set an overall vision and direction for the sector, while not being overly prescriptive in order to allow for greater indigenous technological innovation, adoption, and strategic engagement in setting standards at the international stage. China would benefit from achieving a balance between government regulations and free market dynamics, and between matching the supply and demand of commercially successful applications.

Given the cross-cutting nature of these technologies, progress in the ICT sector will have a significant impact throughout the economy. It is our hope that this report, prepared by a team of the World Bank Global ICT Department in collaboration with many Chinese experts, will contribute to developing a suitable ICT strategy for China, as well as to providing useful insights on how these technologies could best support economic growth, employment creation, and social development.

Mohsen Khalil
Director, Global ICT Department
The World Bank Group

Acknowledgments

This publication, prepared at the request of China's State Council Informatization Office (SCITO) and the Advisory Committee for State Informatization (ACSI), provides a comprehensive overview of the country's information and communication technology (ICT) sector. An earlier draft was submitted to the government in September 2005 as an input to the development of China's 10th Five-Year Plan. The publication was jointly funded by the World Bank and the Department for International Development of the U.K. government.

The publication was written by Christine Zhen-Wei Qiang (Task Manager). In preparing each chapter, she incorporated key written inputs and background papers provided by Professor Gao Xinmin (Vice Chairman, Policy and Planning Committee, ACSI); Ouyang Wu (Director, Policy and Regulation Division, State Council Informatization Office); Yang Yiyong (Deputy Director General, Economic and Social Development Institute, National Development and Reform Commission); Yu Xiaohui (Director, Telecommunications Planning Research Institute, China Academy of Telecommunications Research, Ministry of Information Industry); Zhang Xianghong (Senior Vice President, China Center for Information Industry Development); Zhao Xiao (Director, Macroeconomics Department, Economic Research Center, State Economy and Trade Commission); and Bruno Lanvin, Michael Minges, David Satola, and Randeep Sudan (World Bank).

Overall guidance for the publication was provided by David Dollar (Country Director, China); Mohsen Khalil (Director, Global ICT Department); and Bert Hofman (Lead Economist, China). Special thanks go to Qu Weizhi (Chairwoman, ACSI; then Executive Vice Minister, SCITO) for her important guidance and valuable support. Valuable contributions and comments were provided by Liu He (Vice Minister, Office of the Central Financial and Economic Leading Group); Zhou Hongren (Executive Vice Chairman, ACSI); He Jiacheng (Chairman, Board of Supervisors for Major State-owned Enterprises), Hu Angang (Professor, Tsinghua University); Hou Yongzhi (Senior Researcher, State Council Development Research Center); and Jared Green, Warren Greving, Naomi Halewood, Subramaniam Janakiram, Nikunj Jinsi, Kaoru Kimura, Zaid Safdar, Peter Smith, Jiro Tominaga, Giorgio Valentini, and Bjorn Wellenius (World Bank).

The publication team is grateful to SCITO and ACSI for providing excellent fact-finding, research, and general collaboration, during the 18-month period of the

Acknowledgments

World Bank's study. In particular, the cooperation received from Xu Yu (Secretary General, ACSI); and Fang Xinxin (Deputy Secretary General, ACSI) is sincerely appreciated. This publication incorporates the many valuable comments on earlier drafts received from participants at a May 2005 workshop held in Suzhou, an internal Bank review meeting in March 2006 in Washington, D.C., and at a high-level workshop in May 2006 in Beijing to discuss preliminary findings.

The publication was edited by Paul Holtz. Lansong Zhang, Leona Luo, and Andrea Ruiz-Esparza provided able assistance with administrative and logistical arrangements for the publication team during missions to China as well as follow-up with provincial and central government officials after the missions returned to Beijing and Washington, D.C.

Abbreviations

ADSL	asymmetric digital subscriber line
ADULLACT	Association of Developers and Users of Open Source Software in Administrations and Local Communities
ASEAN	Association of South East Asian Nations
BPM	business process management
B2B	business to business
B2C	business to consumer
B2G	business to government
CAD	computer-aided design
CDMA	code division multiple access
CERNET	China Education and Research Network
CIO	chief information officer
CNNIC	China Internet Network Information Center
CRM	customer resource management
DLD	domestic long distance
DMB	digital multimedia broadcast
DVB-H	digital video broadcast-handheld
ERP	enterprise resource planning
FDI	foreign direct investment
G2B	government to business
G2C	government to citizen
GDP	gross domestic product
GPT	general-purpose technology
GSM	global system for mobile communications
G2G	government to government
IC	integrated circuit
ICT	information and communication technology
ILD	international long distance
IPR	intellectual property right
IPTV	Internet protocol television
ISP	Internet service provider
IT	information technology
ITU	International Telecommunication Union

LAN	local area network
MII	Ministry of Information Industry
MIS	management information system
MOR	Ministry of Railway
MRP	materials requirements planning
NCRE	National Computer Rank Examination
NDRC	National Development and Reform Commission
OA	office automation
OECD	Organisation for Economic Co-operation and Development
OFCOM	Office of Communications
PC	personal computer
PHS	personal handy phone system
PPP	public-private partnership
R&D	research and development
SARFT	State Administration of Radio, Film, and Television
SCDMA	synchronous code division multiple access
SCILG	State Council Informatization Leading Group
SCITO	State Council Informatization Office
SETC	State Economic and Trade Commission
SME	small and medium-size enterprise
SMIC	Semiconductor Manufacturing International Corporation
SMS	short message service
TD-SCDMA	time division-synchronous code division multiple access
3G	third generation
UNCITRAL	United Nations Commission on International Trade Law
VoIP	voice over Internet protocol
VSAT	very small aperture terminals
WAPI	WLAN Authentication and Privacy Infrastructure
W-CDMA	wideband-code division multiple access
WLAN	wireless local area network
WTO	World Trade Organization

Note: All dollar amounts are U.S. dollars unless otherwise indicated.

Overview

Information and communications technology (ICT) is a general-purpose technology (GPT) that can fundamentally restructure an economy. Unlike incremental technical progress, where technological change occurs in small or predictable steps, GPT represents a radical innovation and produces discontinuity in the path of technological development. ICT is an innovation that facilitates and enhances further innovations. It has made product and process innovation much easier—resulting in faster growth in the number of intellectual property rights and patents issued than during any other period in history, as well as a general acceleration of economic processes. In response, the entire economic horizon has changed.

Informatization is not just an economic phenomenon but a social transformation as well and has attributes of a public good. ICT offers value by processing, organizing, storing, and transmitting information. The social effect of ICT is comparable to that of printing: both enable people to gain knowledge at dramatically lower costs. Moreover, information is commodious and ubiquitous, making its distribution costs marginal—while the fixed cost of producing and retaining ownership of information can be very high. This disparity, which can lead to underestimates of the cost of information by the market, may require government involvement to correct market failures.

Informatization is the ICT-driven transformation of an economy and society—not an end in itself but a complex process for achieving more critical development goals. ICT helps countries achieve those goals by spurring innovation, using resources more efficiently, and increasing productivity. Since the 1980s a variety of countries—both developed (Finland, Ireland, Republic of Korea, Norway, Singapore, and Sweden) and developing (Brazil, Chile, Estonia, and India)—have made considerable progress in using ICT and promoting informatization. Success requires supportive policies and regulations, local capacity building, and effective technology implementation and partnerships.

ICT has played a prominent role in China's development strategies since the mid-1990s. China has the world's largest telecommunications market, and its information technology (IT) industry has been an engine of economic growth—growing two to three times faster than gross domestic product over the past 10 years. E-government initiatives (such as the Golden Projects) have achieved significant results. In recent years, the private sector has increasingly used ICT for production and service processes, internal management, and online transactions.

China's development has entered a new stage and requires a new informatization strategy. Industrialization, urbanization, and foreign direct investment are creating unprecedented opportunities—and challenges—for China's informatization. The economy will continue to shift from capital-intensive industry toward information- and technology-intensive manufacturing and services. The evolving needs of firms, particularly those with foreign links, and of the growing middle class are increasing demands for informatization. At the same time, China faces structural barriers that inhibit further informatization. For example, informatization has occurred while China is undergoing more basic industrialization and as it tries to move from a planned socialist economy to a socialist market economy.

China's new informatization strategy should reflect national goals, taking into account the country's stage of development and the economic and social development challenges posed by a large country and large population. The strategy should support:

- Institutional changes to make government functions more service-oriented, efficient, and transparent. Doing so would make markets and resource allocations far more efficient.

- Growth of the services industry (such as IT services and IT-enabled services), including employment opportunities.

- ICT use in reforming manufacturing and energy industries, cutting the cost of capital, and increasing the value added of Chinese products, as well as efforts to increase Chinese enterprises' productivity, international competitiveness, and capacity for technological innovation in a broad range of products and processes.

Informatization Enablers and Building Blocks

Four key enablers and building blocks for achieving these priorities are the legal and regulatory framework, telecommunications infrastructure, ICT industry, and levels of IT literacy and ICT skills.

Establishing an Enabling Legal and Regulatory Environment

Several areas of China's legal framework require high-level legislation in the short and medium term. Although it may be premature to develop an overarching law on

Overview

informatization, legal and regulatory reforms are urgently needed. Areas where legal and regulatory reform can support informatization—telecommunications, network security, Internet content regulation, data protection and privacy, open access to government information, stronger protection for intellectual property rights, efforts to counteract cyber crime—are priorities, as is implementation of the recent E-Signature Law. However, the enabling environment also needs to be improved by amending existing laws.

The complex division of regulatory responsibilities fosters uncertainty. The absence of a legal framework stipulating the principles and scope of informatization makes regulations unclear. Coordination and cooperation among regulatory departments are weak, and China's myriad agencies have different and sometimes overlapping responsibilities. As a result many agencies often regulate the same area—yet accountability is lacking, and resulting regulations are inconsistent, making implementation and execution of laws and regulations problematic. The rule of law should be strengthened to ensure that laws and regulations are enforced and enterprises and government agencies are held accountable.

The legal system should mitigate imbalances in rights and obligations. Many laws and regulations are designed to be administratively convenient, which is often considered more important than individual or enterprise rights. Among enacted laws and regulations, there are far more restrictive provisions than other measures, such as for self-regulation and dispute resolution. Legal obligations are given more weight than rights. An emphasis on processes and procedures (such as licensing) has overlooked protection of privacy and individual rights. Inadequate attention is paid to protecting copyrights and personal data. An enabling legal and regulatory system should focus on encouraging innovation and avoid restrictive provisions on research and development (R&D) that hinder technological change and economic development.

Enhancing Telecommunications Infrastructure

China's telecommunications infrastructure has made rapid progress over the past decade. Since the 1990s China's telecommunications market has become more competitive. As of the end of 2005, there were more than 740 million fixed and mobile telephones (ITU 2006). Close to 50 million computers were connected to the Internet, serving 111 million users—about half of them broadband users (CNNIC 2006). Falling prices for information and communication services and improvements in telecommunications infrastructure provide a solid foundation for further informatization. Still, penetration rates in China remain low relative to Brazil, Russia, other Asian economies, and high-income countries.

Expanding rural ICT access is critical to equitable social and economic development, and China needs to explore ways of narrowing gaps in access. China has not yet established a universal service policy, such as a fund to finance infrastructure expansion in underserved and unserved areas. Nor has it started collecting fees from operators to support infrastructure coverage obligations. In 2004 the Ministry of Information Industry initiated a project to make telephone service available to all villages, requiring the five main telecommunications providers and China Satellite Communications Corporation to share obligations for universal access based on geographic divisions. Subsidies to lower service charges could increase rural penetration. Alternatively, competitive bidding among operators for subsidized contracts would help define obligations and deliver services transparently. In any case, the government should adopt an open policy to allow for technology-neutral ways of implementing universal access to ICT.

Although expanding the broadband Internet market is a clear strategic direction for developing China's telecommunications infrastructure, several barriers impede it. Many potential users remain uneducated about the service or lack the capacity or income to use it. In addition, broadband access depends on personal computer ownership. The government should intensify competition among fixed-line providers, cable television companies, and wireless operators to make broadband costs affordable to more consumers and enterprises.

To support convergence, the government needs to address policy and regulatory issues that prevent broadcast, broadband, and telecom providers from accessing one another's networks. Establishing an independent ICT regulator would be a crucial step to support convergence. Some government reorganization may also be needed where agencies have overlapping responsibilities. Because network mergers involve interconnecting various systems, technologies, and standards, China may need uniform approaches and revised spectrum management policies to support network convergence. Finally, continuous R&D is needed in technologies that enable convergence. To develop needed hardware and supporting applications, China's government, research institutes, operators, equipment manufacturers, and content and service providers need to establish a comprehensive, multistakeholder process to set key convergence goals.

Developing and Innovating the ICT Industry

China's ICT industry has grown rapidly since the mid-1980s thanks to government support for domestic companies and R&D. Both the central and the local governments are promoting the industry by facilitating funding for startups and incubators. In addition, the government has provided incentives for foreign investment—while also requiring foreign firms to transfer technology in return for market access. Given the size and potential of China's market, many foreign firms have been willing to make this tradeoff. China led the world with $180 billion in ICT exports in 2004, surpassing those of the United States.

A major challenge for China's integrated circuit and computer industries is to move beyond production of low-end products and applications, climb the value chain, and expand to the global market. Although China is a leading exporter, it still has to import chips to meet local needs. About 90 percent of the chips manufactured in China are exported, yet a large share is re-imported after processing. This means that foreign firms add value to chips produced domestically that are ultimately consumed domestically—highlighting the low end of the value chain where China's integrated circuit industry is positioned. Given the country's strong technical skills and ability to attract foreign investment and technology, it clearly has the potential to move up this chain.

Three notable features of China's software industry are its product focus, regional dispersion, and lack of pure-play outsourcers. Many software companies with strong capacities and business models, often in regions with extensive high-technology infrastructure, have a product orientation—leaving IT services undeveloped. The outsourcing industry is highly fragmented and lacks large players dedicated to outsourcing. Recently, however, foreign companies moving their software development operations to China (although to serve their corporate parents rather than third party customers) have become a key driver in the country's outsourcing market. Such centers create new software export markets and make China a major link in the global software value chain.

Chinese IT service firms face low domestic demand, extensive piracy, and intense intellectual property rights challenges, giving firms little incentive to invest in product development. Although China aims to develop an IT services industry that works for foreign firms, many Chinese companies lack English language skills and have no experience in U.S. or other foreign markets. Until China develops the human capital required to expand this industry, the country will remain a low-cost location for coding and maintenance and is unlikely to create a software industry that can rival Indian (among other) giants.

Digital media could become a significant industry in China, but its development is hindered by weak R&D, a shortage of developers, and tough restrictive regulations on digital content. Given the looming introduction of a third-generation network and the growing integration of ICT and traditional content industries, digital media has developed a large service market in a short period, jumping to more than $12 billion in 2005. The industry is expected to maintain high growth over the next 5 to 10 years. However, government policies on digital content can be contradictory, promoting the industry's development while also asserting stringent regulation. This is partly due to the potential negative impact of digital media.

Stimulating innovation and supporting R&D are essential for the ICT industry to attract investment, maintain high growth, and become globally competitive. The demand for ICT from domestic and foreign markets creates opportunities and incentives for investing in additional R&D. First, however, China must ease the obstacles to a more effective national innovation system by removing regulatory obstacles to the introduction of new technologies, eliminating entry barriers and allowing more foreign direct investment, and aligning standards development with international practice.

China should establish strategic directions around core technologies and improve collaboration between academia and businesses to focus R&D on relevant applications. Core technologies, products, and services—particularly those unique to China's market or where it has strategic advantages—are critical to the country's industrial development and competitiveness. These include integrated circuits, network security software, telecommunications equipment, and mobile data applications. China needs to further integrate enterprises, universities, and research institutes to connect skills development and R&D with industrial development, and to strengthen links between production and demand—especially those from the domestic market.

Improving ICT Human Resources

Three levels of human resources are critical to informatization: a general public able to use ICT applications at work and home; informatization managers who lead ICT development in government and business; and ICT professionals experienced in network design, software development, and R&D. Stakeholders—governments, enterprises, schools, research institutes, and individuals—must work together to build the human resources needed to maximize the economic and social benefits from ICT development.

China has a major shortage of the skilled ICT workers needed to implement its informatization strategy and maximize the strategy's economic impact. The 4 million workers in China's IT industry in 2003 accounted for less than 1 percent of the country's labor force. It is essential to quickly develop human resources of professional caliber at various levels, including ICT professionals with multidisciplinary management skills. A variety of education and training programs—including links between universities and businesses—can help achieve this goal.

The brain drain of skilled workers to more developed regions and countries should be mitigated as much as possible. Limited awareness of the significance of informatization for economic development leads to low investment in ICT infrastructure, impeding investment in ICT programs. All these factors drive away educated, technologically savvy residents of underdeveloped areas. Thus it is essential to raise awareness of the importance of informatization and encourage greater public participation. At the same time, ICT diversifies educational opportunities by overcoming shortages of teachers and classrooms in remote areas through distance learning (delivered by radio, television, or online).

ICT Applications

These enablers, with the support of solid leadership, should aim to transform administrative and business processes into automated, streamlined processes that support the two pillars of ICT applications: e-government and e-business.

Advancing E-Government

China's government has played a significant role in stimulating ICT demand and supply. The government is the country's largest investor in ICT and leads the adoption and use of ICT applications. China began incorporating ICT networks and applications into government processes in the mid-1980s, with government departments generally adopting internal informatization appropriate to their functional goals. In recent years, the central government has become more ICT-capable through Golden Projects—initiatives designed to make public services more efficient and transparent. E-community applications at the local government level, such as call centers and Web sites, provide citizens with direct access to public services and improve information flows, strengthening interactions between officials and citizens. Rural areas in particular can benefit from online information services: access to relevant information can transform economic opportunities and improve livelihoods for rural households.

Many investments in large-scale informatization have had mixed results. E-government applications are often huge management information system projects requiring large investments. At the same time, one of the goals of e-government is to reduce the transaction costs of government operations. Thus it would be useful to adopt a clear methodology for prioritizing e-government investments and maximizing returns. For example, Australia uses a demand-and-value assessment methodology to determine which e-government applications deserve funding. The Chinese government could consider such an approach to ensure that e-government investments yield tangible results.

Detailed feasibility studies should be conducted before large projects are undertaken. ICT suppliers in China often build networks, hardware, and software based on existing technologies and products—failing to achieve the aims of government agencies to re-engineer administrative and business processes, or to meet the needs of end users such as businesses and citizens. Thus any feasibility study should let practicality guide implementation, based on demand from users. Also crucial are monitoring and evaluation frameworks with clear and measurable output and outcome indicators for each project. In addition, third-party supervision can be used to ensure the quality of e-government projects.

Mechanisms are needed for integrating and sharing information resources. Gaps in economic development have generated huge differences among Chinese regions in information resource development and use. Sharing and exchanging information resources could be helpful, especially for regions and provinces lagging in e-government development. Ireland's Local Government Computer Services Board, the United Kingdom's Local Authority Software Consortium, and the European Union's e-Government Observatory are emerging models for the Chinese government to consider. Although an entrenched culture of secrecy impedes the free flow and sharing of information, provisions on open government information (now in draft form) will likely improve information access.

Fostering E-Business

Competition among foreign and domestic firms will increase the demand for informatization and boost domestic firms' productivity and efficiency. As China begins fulfilling the terms of its accession to the World Trade Organization by further opening its economy, its firms will need to become more competitive to thrive in both domestic and foreign markets. Many Chinese firms recognize the benefits that informatization can have for their operations, and ICT investments have grown steadily in recent years—especially among wholesale and retail firms. ICT applications have significantly improved production and management systems in China's industrial firms, and they play an increasingly important role in the growth of these firms. Among firms that have adopted ICT applications, most report that the effect on their operations has met or exceeded expectations.

A few large companies have led the way in incorporating ICT in their purchasing, advertising, and marketing procedures, while common e-business platforms have become popular for small and medium-size enterprises (SMEs). Business-to-business (B2B) e-commerce systems are often among the most sophisticated Internet business models because they require large amounts of data and commodity exchanges, and so require large Web capacity and office automation and management information systems to manage information inflows and outflows. Common B2B platforms for SMEs enable these firms to join and search for potential clients and suppliers in a sort of commercial information exchange. Meanwhile, business-to-consumer e-commerce, about 5 percent of the e-business market, is a key aspect of informatization for Chinese firms that sell products to consumers.

The government can encourage e-business through national and local initiatives. Government provision of online information and services can demonstrate the potential benefits of ICT to businesses, and can help build trust in the efficacy and security in online transactions. As model users, the central and local governments can also set standards for ICT adoption by firms. To foster access to public services and meet requirements for business purposes, firms should be encouraged to adopt systems and software compatible with e-government services—such as public e-procurement.

Many SMEs that do not have the means to access ICT applications or expertise can be aided by government-sponsored incubation. Most of China's 8 million SMEs are still in the early stages of deploying ICT applications. Firms that want to pursue e-commerce but do not have the means to do so can be assisted through government-sponsored e-business platforms. Such initiatives could be coupled with promotional programs and e-commerce trade shows that demonstrate how e-commerce works and the steps needed for online transactions. Promotional programs could also encourage SMEs that do not export to do so through e-commerce.

Summing Up

Despite significant progress, obstacles remain to accelerating informatization in China. To a large extent, these are systemic problems cutting across the economy and society, tied to China's ongoing economic and social transformation. Addressing all the critical factors is complex and requires long-term commitment. However, several key issues need to be addressed decisively in the second half of this decade, through

Overview

policies entailing institutional reform, to trigger broader changes. They are as follows:

- Fostering indigenous innovation for domestic and foreign markets
- Promoting strategic engagement in setting standards
- Matching the supply and demand of commercially successful applications
- Striking a balance between government regulations and free market dynamics.

Government decisions about ICT can also be seen as decisions on the course of the economy as a whole. In some ways, the problems affecting China's ICT policies and strategies are not much different from those that the country will face in other sectors. However, the rapid pace of technology development means that ICT issues are being addressed before other problems. Moreover, the effects of ICT development will be felt throughout the entire economy.

Chapter 1

China's Emerging Informatization Strategy

Over the past decade, China has devoted considerable resources to developing its informatization strategy and has conducted extensive research to determine the appropriate framework and goals for building an information society. China's informatization has been guided by the Outline of the National Informatization Development Plan (1997) and the Special Planning of Informatization document of the 10th Five-Year Plan (2000). As the strategy nears its 10-year mark, it requires updating to reflect the evolving needs of China's economy.

The revised informatization strategy will require a multistakeholder approach that encourages wider use of information and communication technology (ICT) to foster economic and social development. Progress on informatization will depend heavily on China's enabling environment, including its legal and regulatory framework, telecommunications infrastructure, ICT industry, and human resources. Features of ICT (box 1.1), global trends, and lessons from China's experiences with informatization can provide direction and a general framework for the new strategy.

Informatization and Development: The Global Picture

Informatization is the ICT-driven transformation of an economy and society—a complex development process in which a country increases its capacity to exchange and apply information and, in turn, generate knowledge. The informatization process involves investing significantly in infrastructure that facilitates the use of ICT by government, industry, and the general public.

Informatization is not an end in itself but a process for achieving more critical development goals. ICT helps countries achieve those goals in several ways:

- First, it spurs innovation. ICT—including the Internet—allows information and knowledge to be shared more easily, facilitating new forms of economic and social interaction.

Box 1.1 Characteristics of Information and Communication Technology

How does informatization compare with previous waves of technological innovation in China and elsewhere? Answering that question requires understanding several features of ICT.

- First, ICT is a general purpose technology that establishes a new technological paradigm and results in a fundamental restructuring of the economy in general, and of production in particular. Unlike incremental technical progress, where change occurs in small or predictable steps, general purpose technologies represent a radical innovation and produce discontinuity in the path of technological development—changing the direction of development more broadly. Thus informatization is not just a technological phenomenon but an economic and social transformation as well.

  The remarkable contribution of ICT to total factor productivity has been demonstrated in numerous studies (summarized in Qiang and Pitt 2003). While ICT has created new industries and helped transform firms, its economic and social effects are complicated. Furthermore, general purpose technology innovations—defined as those with large, extensive, and prolonged economic impacts, such as the steam engine and electricity—typically take a long time to permeate through and significantly affect the overall economy, as David (1991) documents for developed countries. It may take even longer for general purpose technology to spread in developing countries, due to lower levels of technological knowledge. At the same time, the relative lack of entrenched technological structures may enhance these countries' potential for leapfrogging. Thus the role of ICT in transforming the economies and societies of developing countries should be thoroughly analyzed and understood.

- Second, ICT is an innovation that enhances further innovation methods and processes. ICT is widely used to gather and transmit information, design complex new products, and coordinate and conduct research and development in different areas, including marketing. This has made product and process innovation much easier, resulting in faster growth in the number of intellectual property rights and patents issued than at any other period in history, as well as a general acceleration of economic processes. In response, the whole economic horizon has changed.

  Around the world, ICT has changed the structure of manufacturing and service industries and of production chains. Routine activities—such as manufacturing, assembling, basic administration, and customer service—are increasingly being undertaken in developing countries, while creative and highly skilled activities—such as product design, business strategizing, workflow design, marketing, and management—remain in developed countries. This production model has become especially common for digital products. Furthermore, the widespread use of broadband technology has enabled the standardization and outsourcing of white-collar jobs and business workflows.

- Third, information and knowledge have attributes of public goods and so may require government involvement to correct market failures. Unlike other general-purpose technologies, ICT can become a source of economic value by processing, organizing, storing, and transmitting information. The social effect of ICT is comparable to that of printing, both of which enable people to gain knowledge at dramatically lower costs.

(Continued)

China's Emerging Informatization Strategy

> **Box 1.1** **(Continued)**
>
> Moreover, information is commodious and ubiquitous, making its distribution costs marginal. The fixed cost of producing and retaining ownership of information can be very high. This means that the market may underestimate the cost of information; accordingly, relying on the market alone could lead to an inadequate supply of information, falling short of demand. Thus government intervention may be needed to make up for the short supply.

- Second, ICT supports the efficient use of economic resources. For example, informatization requires fewer natural resources than do traditional processes of development, such as manufacturing and industrialization. ICT also supports sustainable economic growth and the development of postindustrial service economies.

- Third, ICT increases productivity and thus international competitiveness. Technological innovation is a central determinant of competitiveness in the global economy, and a country's informatization strategy can help develop an enabling environment for it.

Expanding ICT use is crucial to modernization, and since the 1980s all but the poorest countries have increasingly applied ICT to achieve economic and social goals. A variety of countries—both developed (Finland, Ireland, Republic of Korea, Norway, Singapore, and Sweden) and developing (Brazil, Chile, Estonia, and India)—have made considerable progress in promoting informatization and fostering enabling environments for new technology.

A recent World Bank study of 40 national ICT strategies shows that countries have pursued similar informatization paths (figure 1.1), with ICT development often occurring in three stages:

- *First stage.* The focus is on building infrastructure and developing technology, with a relatively fragmented approach to applying ICT in the economy, government, and society.

- *Second stage.* The process develops a unifying policy vision that relates ICT to overall economic and social development and provides a basis for coordinated action across these three different policy areas: economy, government, and society.

- *Third stage.* ICT initiatives are coordinated at the policy level and implemented in a way that is mutually reinforcing, leading to the transformation of industrial, economic, and social structures (World Bank 2006c).

The positions of countries on this continuum of ICT policy and strategy development generally correlate with their income levels. Most developing countries are improving their infrastructure and trying to move from the first stage to the second. Developed countries are beginning to move to the third stage.

Figure 1.1 General Pattern of Informatization Strategy Development

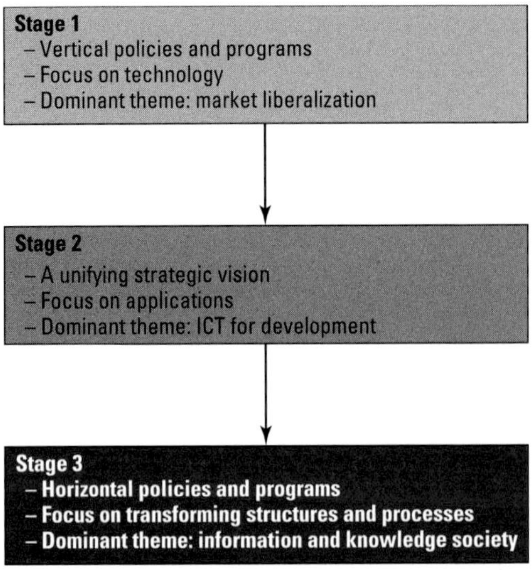

Stage 1
– Vertical policies and programs
– Focus on technology
– Dominant theme: market liberalization

↓

Stage 2
– A unifying strategic vision
– Focus on applications
– Dominant theme: ICT for development

↓

Stage 3
– Horizontal policies and programs
– Focus on transforming structures and processes
– Dominant theme: information and knowledge society

Source: Author.

ICT development involves numerous challenges at each stage. China's informatization strategy should draw on experiences elsewhere that show success requires the following:

- *Effective technology implementation.* Informatization requires fostering the development of a wide range of technologies—from leading-edge Internet and telecommunications infrastructure to new applications for commerce, trade, health, education, agriculture, government, and online security—and achieving high levels of ICT infrastructure and access to narrow the digital divide.

- *A supportive enabling environment.* Governments must develop policies and regulations that stimulate ICT deployment, taking into account—and where necessary, adjusting—political, economic, and legal systems (including the organizational structure of government bodies involved in ICT policy making) as well as the level of public awareness of ICT's role in development.

- *Partnerships and alliances.* To develop projects that benefit all stakeholders, informatization efforts must include private enterprises, universities, and research and development (R&D) institutes.

- *Local capacity building.* Communities must be given support, through human resource development, to develop and manage their own ICT projects.

Although many informatization strategies are linked to economic growth objectives, few are aimed directly at alleviating poverty. Moreover, many countries' ICT strategies suffer from a lack of coordinated management, monitoring and evaluation mechanisms, and efforts to address digital divide issues. Given China's vast size and

China's Emerging Informatization Strategy

population, an effective implementation strategy will be critical to ensuring that its informatization strategy serves its development goals.

Informatization in China over the Past Decade

Since the mid-1990s strong leadership at the national, provincial, and municipal levels, combined with a sustained, long-term focus on ICT development at the highest levels of the Chinese government, has given ICT a prominent role in sustainable development strategies and ensured its application to core economic and social goals.

In 1993 China's State Council established a high-level National Economic Information Joint Committee. The committee launched the first-tier and second-tier Golden Projects to promote informatization focused on economic goals; in 2002 these were followed by the sectoral Golden Projects (see chapter 6), which have laid a solid foundation for further e-government and e-commerce development. In 1997 the Outline of the National Informatization Development Plan established a framework for promoting informatization, supporting the development of related policies and regulations, information resources (content), network infrastructure, applications, the information technology (IT) industry, and human resources. In 2000 the Fifth Plenary Session of the 15th Central Committee of the Communist Party of China established a strategy of "promoting industrialization through informatization" (Qu 2005). In 2001 the State Council Informatization Office, under the State Council Informatization Leading Group, headed by the premier, was created to generate policy proposals, coordinate strategy implementation, draft laws and regulations, set standards, and develop plans for China's information security. Guidelines to promote e-government and support the development of China's software industry were also issued.

China's policy and legal environment has been adapted to promote informatization. A telecommunications act and regulations on government information publicity have been drafted and are being reviewed. The E-Signature Law went into effect in April 2005 (Letner 2005). In addition, many provincial and municipal governments have dedicated resources to coordinating and managing ICT initiatives.

As a result China has made enormous progress in developing its telecommunications infrastructure and IT industry—essential components of an information society. In response to policy reform, deregulation, and liberalization of the telecommunications market—which has introduced competition among fixed and mobile telephone and Internet service providers—telecom infrastructure has expanded rapidly over the past decade. There were more than 740 million fixed and mobile telephones by the end of 2005, or around 50 per 100 inhabitants, according to the National Bureau of Statistics (NBS 2005). Close to 50 million computers were connected to the Internet, serving 111 million users—including more than 64 million broadband users (CNNIC 2006).[1] Falling prices for information and communication services and improvements in telecommunications infrastructure have provided a solid foundation for promoting further informatization.

China's IT industry has been an engine of economic growth, expanding two to three times faster than the gross domestic product (GDP) over the past 10 years. The industry accounts for more than 15 percent of total GDP growth, and in 2004 it represented 7.5 percent of GDP—up from 4 percent in 2001 (Qu 2005). Moreover, IT products account for 28 percent of China's exports—electronics companies such as Huawei and ZTE are now major multinationals.

China's progress on informatization is also evident from e-government applications in both central and local governments. E-government initiatives have achieved significant initial results. The e-government system developed for public revenues and expenditures—known as the Golden Tax (e-taxation) project—has enabled value added taxes to be processed and tax invoices audited through Web-based applications, improving tax collection and increasing tax revenues. Tax administration authorities in medium-size and large cities now allow enterprises to pay taxes online. In addition, the Golden Gate (e-customs and foreign trade) application has helped reduce smuggling and fraud by curtailing false customs declarations. The Golden Wealth (e-fiscal management) initiative has improved the capacity of the treasury payments system to supervise the use of public financial funds.

China's e-government efforts have also contributed to the spread of ICT, particularly in urban areas. More than 90 percent of municipal governments have established Web sites, and many large cities have developed applications that enable online interactions and business transactions. Examples include Beijing's "review and approval" system, Shanghai's social security card system, Guangzhou's community service platform, and the integrated emergency response system of Nanning City in Guangxi province.

Government-run ICT applications and infrastructure have also supported rural development. A Web site has been established that collects and provides daily price information for 280 large wholesale agricultural markets and more than 300 agricultural products. In addition, 9,000 villages and towns have been connected to the Internet, raising the coverage of villages to 23 percent. The Golden Agriculture project, soon to be initiated, will serve as a key national ICT initiative.

In recent years, the private sector has increasingly used ICT for online business-to-business and business-to-consumer transactions, contributing to growth in e-commerce. Many of China's financial and banking institutions now transfer funds through secure online systems. E-payment applications, involving bank cards or online payment systems, have also been developed. In 2002 more than 500 million bank cards were issued, worth more than $1.2 trillion. The Silver United Card, a credit card that works with a network of financial institutions, is available in 348 cities.

Some 80 percent of medium-size and large enterprises now conduct online transactions in some form. Telecommunications, civil aviation, petrochemical, power, and manufacturing firms have automated production and service processes and applied IT to internal management processes. Some large enterprises, such as the China National Petroleum Corporation and China Petroleum and Chemical Corporation have also developed online purchasing and e-commerce platforms with positive initial results.

China's Emerging Informatization Strategy

Informatization and Economic Development in China

Since its economic opening more than 25 years ago, China's development has entered a new stage and requires a new informatization strategy. Informatization and economic development are mutually reinforcing: China's successful development has laid the material and technological foundations for informatization while greatly expanding the demand for it. Urbanization, industrialization, upgraded consumption, and increased social mobility have created unprecedented opportunities as well as new challenges for the informatization process. China's future growth will likely come from three sources:

- encouraging services and domestic demand
- improving firm competitiveness
- facilitating the movement of labor out of agriculture and to the cities.

To sustain the momentum of economic growth, China will need to closely examine three interwoven factors during the 11th Five-Year Plan (2006–2010) and well into the future.

Economic Structure

Industrialization has resulted in continuous changes to China's economic structure. Many people are shifting from agricultural to industrial and service jobs with higher value added (figure 1.2). Electronic communications, real estate, and social services are among the fastest-growing sectors, in terms of both value added and employment.

Productivity has increased considerably in industries that produce ICT-related equipment as well as among ICT-using providers of wholesale, retail, financial, and logistics services (World Bank 2004). These increases suggest that the economy will

Figure 1.2 China's Economic Structure by Sector, 1978–2003

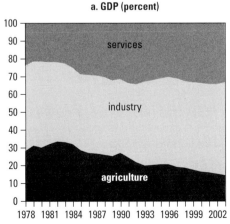
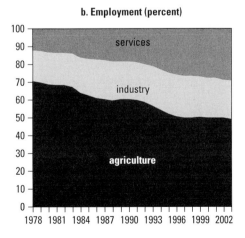

Source: National Bureau of Statistics 2004.

Figure 1.3 Foreign Direct Investment in China, 1996–2005

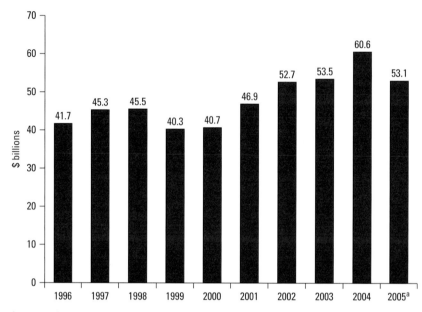

Source: UNCTAD 2006.
a. The data are for January through November.

continue to shift toward information- and technology-intensive manufacturing and service sectors.

Foreign Direct Investment

Foreign direct investment (FDI) continues to flow into China at a growing rate, mainly to develop manufacturing capacity for exports but increasingly to service domestic demand. In 2004 China attracted $61 billion of FDI—up 13 percent from the year before (figure 1.3). In the decade from 1996 to 2005, China accumulated $480 billion of FDI.

Foreign firms invest in China primarily to take advantage of its low labor costs. However, an increasingly important benefit for China is the opportunity to transfer R&D efforts, through new technology embedded in equipment and processes, demonstration effects, circulation of managers and workers, competitive pressures, and technological links among foreign firms and their suppliers and subsidiaries. All of these effects provide incentives for businesses to adopt ICT. Aside from contributing to productivity, the spread of ICT has the potential to accelerate innovation in a broad range of products and processes. Foreign firms also use informatization to penetrate the domestic market.

Urban centers in Guandong, Jiangsu, Fujian, and Shanghai absorb more than half of China's FDI, reflecting their strong base of skills, coastal locations, fewer regulatory impediments, and proximity to two of the country's biggest investors, Hong Kong (China) and Taiwan (China). The geographic distribution of FDI in

China's Emerging Informatization Strategy

China is building technological capability and augmenting agglomeration effects in coastal metropolitan areas.

Productivity and innovation at the firm level are functions of ICT use and FDI. Companies that use ICT grow faster, employ more skilled workers, and are more productive and profitable than those that do not (Qiang, Clarke, and Halewood 2006). Among Chinese firms, joint ventures and firms linked to international production networks invest most heavily in IT, highlighting the significance of FDI (Yusuf and Evenett 2002).

Urbanization and a Growing Rural-Urban Income Divide

Since the 1980s, China's growth has been driven by urban areas and their industrial activities. Accordingly, the country has experienced rapid urbanization, with the urban share of the population reaching nearly 40 percent in 2004, up from 30 percent in the 1990s (figure 1.4). Rural emigration has accounted for nearly 70 percent of the increase in the urban population.

The fastest-growing industries are in major cities on the coast as well as a few along main inland waterways. Among provinces and cities with provincial status, immigration has been highest in Guangdong, Shandong, Shanghai, Beijing, and Jiangsu. In 2002, 482 million people lived in coastal provinces, and 425 million in the central region (World Bank 2004).

The income gap between urban and rural population has widened. By the third quarter of 2005, average per capita urban income was 3.26 times as much as the rural average income (*People's Daily* 2005). The growing gap is partially a result of China's development strategy focusing on industrialization, which tends to be urban-centered.

Figure 1.4 China's Urban-Rural Population, 1995–2004, and Income Ratio, 1978–2004

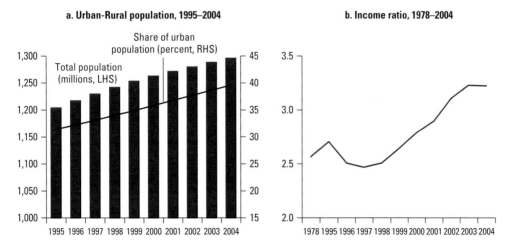

Sources: Huang and Pieke 2003; Park 2004.
Note: LHS = left-hand side;
RHS = right-hand side

The evolving needs of firms based in coastal areas—and of the growing middle-class concentrated in cities in those areas—are increasing the demand for informatization. To support informatization and sustain growth, coastal areas need to develop high-quality infrastructure—including access to ICT facilities and services—for the growing urban centers that drive economic performance and serve as the locus of innovation. However, provinces seeking to catch up also need to grasp informatization so as not to intensify the existing urban-rural economic and social gaps.

A Framework for China's New Informatization Strategy and Challenges

According to the State Council's Development Research Center, between 2005 and 2020 China's economy will grow by 7 percent to 8 percent a year (Lu 2005). By 2020 per capita GDP is expected to reach $3,000, keeping the country in the lower-middle-income group. Thus China's new informatization strategy should be put into the context of this development level, rooted in a realistic understanding of objectives and approaches, and taking into account the asymmetric economic and social development challenges posed by a large country and a large population.

The overarching, long-term goal of informatization is to support China to achieve rapid sustainable economic growth that benefits all citizens. Over time China will shift from a growth pattern that favors capital-intensive industry to one that is less resource-intensive, more efficient, and more equitably shared. That will involve rebalancing growth to sectors that require less capital, energy, and resources, and generate more employment. Thus informatization could support the following (figure 1.5):

- institutional changes of making government functions more service-oriented, efficient, and transparent, contributing enormously to market efficiency and resource allocation
- growth of the service industry (such as IT services and IT-enabled services) and employment opportunities
- ICT use in reforming manufacturing and energy industries, reducing the cost of capital, and increasing the value added of Chinese products
- efforts to increase Chinese enterprises' productivity, international competitiveness, and capacity for technological innovation in a broad range of products and processes.

For the overall informatization strategy to progress, the framework in figure 1.5 links these priorities and goals to drivers and enablers. This report identifies four key enablers and building blocks to achieve these objectives:

- legal and regulatory framework
- telecommunications infrastructure
- ICT industry
- IT literacy and ICT-skilled labor.

China's Emerging Informatization Strategy

Figure 1.5 Framework for China's Informatization

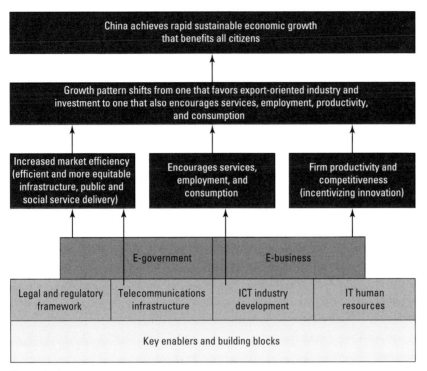

Source: Author.

These enablers, with the support of solid leadership, should aim to transform administrative and business processes into automated, streamlined processes that support the two pillars of ICT applications: e-government and e-business.

Several challenges in the enabling factors, if not addressed properly by the new strategy, would impede continued progress on China's informatization:

- *The institutional setup of ICT policies and regulations continues to constrain China's informatization.* An enabling environment for ICT development needs to be created, with a host of changes in policies and regulations as well as a reorganization of government agencies involved in ICT policy making. Informatization requires a legal system that can protect identities, intellectual property, and online transactions. In addition, China lacks a coordinating authority that can prioritize ICT investments and manage projects at the national level. Inadequate coordination and persistent interagency conflicts have contributed to redundant and overlapping initiatives.

- *The digital divide in China may intensify the rising inequalities in income and opportunities* that have come partly as the by-product of its transformation from an agricultural to an industrial economy. There are gaps in access to ICT infrastructure between rural and urban areas, as well as between poorer, more isolated western provinces and wealthier, more developed eastern ones. A number of groups remain underserved and have limited access to ICT. China's many minority groups—rural farmers, the vast "floating" population, the elderly, and the learning disabled—have increasingly less opportunity to join the information society.

- *China relies heavily on foreign technology and expertise,* which makes ICT equipment and services more expensive and therefore less accessible. A major challenge for China will be to expand beyond production for the domestic market and low-value peripherals and components for the global market. Its capacity for R&D, product development, and ICT innovation are critical for the industry to attract investment, maintain high growth, and become globally competitive.

- *ICT skills are not strong enough to reap the benefits of informatization.* China's competitive edge will be determined by its population's ability to acquire, share, and use information and communication effectively. As informatization accelerates and the public becomes more aware of its importance for China's economic and social transformation, the country will need more ICT experts, as well as a labor force with adequate ICT skills. It has become a pressing task to improve and expand the human resources to meet the demand for China's ICT development.

- With respect to ICT applications, *large-scale informatization projects have often yielded limited results.* Investment returns have been uneven in some areas, and despite high investment in telecommunications infrastructure, demand for ICT applications overall has been relatively weak. Low Internet use and limited demand for applications that can serve China's economic and social development goals suggest underuse of existing infrastructure and an inability to integrate information resources. In addition, IT systems are often improperly managed and maintained after initial implementation.

China also faces fundamental structural barriers that inhibit further informatization. For example, informatization has been initiated while China is undergoing a more basic process of industrialization and as it attempts to move from a planned socialist economy to a socialist market economy. In addition, China's vast, diverse territory and large population increase the difficulty of spreading ICT equitably.

The issues surrounding informatization not only include economic strength and technological advancement, but also social equality and institutional capacity building. As its society becomes more mobile and information-intensive, the Chinese government needs to improve regulations on information flows, overcome a general unwillingness to share information, and foster community development and mitigate social conflicts.

Note

1. Broadband provides high-speed Internet access and enables the development of advanced online applications and content. Broadband networks can also be used to provide Voice over Internet Protocol (VoIP) service (also known as Internet telephone).

Chapter 2

Establishing an Enabling Legal and Regulatory Environment

A key element in achieving the full potential of the contribution of ICT to economic and social goals is a comprehensive and supportive environment. In particular, fostering ICT use and ICT-led growth requires implementing credible, transparent, and nondiscriminatory policies, laws, and regulations. The fast-changing nature of ICT, including the convergence of telecommunications and information technologies, has posed challenges to some of China's policies, laws, and regulations that may no longer fit the new environment.

ICT Regulation and Regulatory Institutions

Currently, the State Council Informatization Leading Group (SCILG) formulates national ICT policies and ensures that these policies serve state interests. But China does not have an overarching institutional framework for informatization that establishes clear responsibilities for the many agencies involved in ICT regulation. As a result government ministries and regulators often have overlapping and conflicting activities, contributing to uncertainty as well as to ineffective regulation and enforcement, and inhibiting private investment and ICT development. In terms of the Internet and its manifold applications, China's regulatory approach, as complex as its institutional arrangements, can be characterized as having two seemingly unrelated tracks.

On the one hand, China is moving toward an enabling environment conducive to economic growth. Increasingly, this enabling environment is embracing good international practices to promote a sophisticated information society, including incremental steps toward protecting intellectual property rights, for example. On the other hand, China's lack of adequate privacy protection and its approach to certain aspects of openness of information are often seen as inhibitory. It remains

Figure 2.1 Institutional Structure of China's Telecommunications Sector

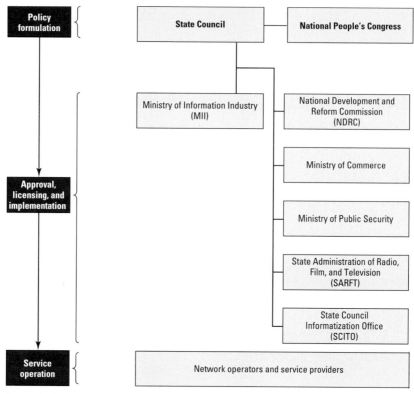

Source: Author.

to be seen what effect the regulations will have on economic and commercial development.

The ICT sector, including telecommunications, broadcasting, and data networks and services, is regulated mainly by the Ministry of Information Industry (MII) and the State Administration of Radio, Film, and Television (SARFT). Numerous other agencies, however, are involved in ICT regulation (figure 2.1; table 2.1).

The MII historically has been the main regulator of telecommunications. Its responsibilities include developing policy and regulation, overseeing the operation of all networks (except cable television), implementing policies approved by the State Council, and setting plans and targets for telecommunications infrastructure development. It is also largely responsible for regulating the Internet.

SARFT regulates broadcasting networks and television administration, defined as facilities that carry television and radio programs—including standard television and radio networks and their relay stations, satellite television and radio networks and their upload and relay stations, microwave stations, and cable television and radio stations.

Establishing an Enabling Legal and Regulatory Environment

Table 2.1 ICT-Related Regulatory Responsibilities of Selected Government Agencies

Agency	Responsibilities
Ministry of Commerce	• Oversees foreign investment, including in telecommunications
Ministry of Information Industry	• Regulates telecommunications and Internet
Ministry of Public Security	• Enforces commercial and criminal laws related to telecommunications networks
National Development and Reform Commission (NDRC)	• Formulates macroeconomic plans • Budgets for and finances telecommunications infrastructure projects
National People's Congress	• Approves laws
State Administration of Radio, Film, and Television (SARFT)	• Regulates broadcasting industry • Administers radio spectrum for broadcasting
State Council Informatization Office (SCITO)	• Regulates Internet content and other media
State Council Informatization Leading Group (SCILG)	• Formulates and coordinates national ICT policies

Source: Author.

The National Development and Reform Commission oversees funding and operations of state enterprises. The State Council also issues regulations governing Internet providers and computer networks connected to the Internet.

Recent advances have given rise to the convergence of telecommunications, broadcasting, and Internet technologies, despite their entirely different histories of ownership, control, and regulation. Each technology also has the ability—and commercial motivation—to encroach on the markets of the others. Responses to these developments have varied. For instance, SARFT has issued 66 licenses for video-on-demand services, but only one has been granted to a telecommunications operator. The rest have been issued to broadcast networks. At the same time, the ability of cable television operators to enter into telecommunications services remains uncertain.

In the absence of regulatory convergence, the division of regulatory responsibilities is vague. Multiple agencies regulate the same technologically converging areas, such as data, Voice over Internet Protocol, and multimedia (figure 2.2), and coordination and cooperation among them are minimal.

Overlapping regulatory authority also occurs in other ICT sectors. For example, the Ministry of Culture and the General Administration of Press and Publication have both claimed regulatory responsibilities over online gaming.

Proposed Framework for China's Enabling Environment

Given the complex and cross-cutting nature of ICT, creating the right environment for its development is a daunting task for policy makers. Although best practices are emerging from countries that have crafted policies and implementation strategies to

Figure 2.2 Regulatory Jurisdictions for China's ICT Services and Networks

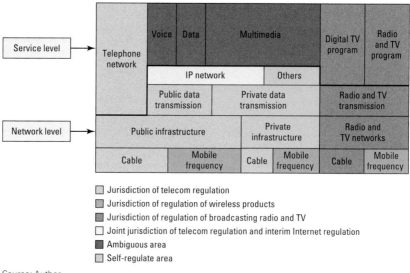

- Jurisdiction of telecom regulation
- Jurisdiction of regulation of wireless products
- Jurisdiction of regulation of broadcasting radio and TV
- Joint jurisdiction of telecom regulation and interim Internet regulation
- Ambiguous area
- Self-regulate area

Source: Author.
Note: IP = Internet Protocol.

facilitate digital opportunities, no universal blueprint exists. Successful reforms must take into account the need for comprehensive changes that cut across traditional technological and commercial boundaries.

In addition, the role of regulators and regulation itself must be reevaluated (Guermazi and Satola 2005). For example, to promote the use of ICT and the Internet for commerce among citizens, business, and government, a framework that grants legal validity to electronic contracts and digital signatures is essential—but it is not sufficient to enable e-commerce on its own. To encourage e-commerce in financial markets, a robust payments system must also be in place. In addition, for any kind of e-commerce, network, and critical infrastructure, the data of users and the intellectual property contained in the hardware and software that drive these activities need adequate security and protection.

The main goal of regulatory reform is to create a stable, open environment that fosters confidence in the ICT market. A major step toward that goal involves establishing clear, transparent governance structures and respect for the rule of law. Basic principles for regulatory reform include encouraging market-based approaches and facilitating market entry, promoting business confidence and clarity, strengthening contract enforcement, ensuring interoperability (for example, of systems, standards, and networks), and protecting intellectual property and consumer rights. Moreover, all regulatory policies should be neutral regarding both the use and the type of technology. Based on international experiences, the proposed enabling environment for China consists of public policies crucial to three main areas of ICT development:

Establishing an Enabling Legal and Regulatory Environment

Table 2.2 Main Areas, Goals, and Policy, Legal, and Regulatory Issues for Informatization

Area	Goal	Policy, legal, and regulatory issues
Infrastructure	Improve access to information and communication infrastructure	• Enacting telecommunications law • Promoting private and foreign investment • Introducing competition regulation • Strengthening institutional capacity • Emerging regulations on broadband deployment and converged services • Addressing access gap
Applications (supply side)	Promote e-applications, including e-government and e-business	• Opening access to government information • Implementing e-transactions (e-signatures and e-contracts) • Promoting Internet content regulation • Promoting online payment regulation • Promoting online taxation
Consumer confidence (demand side)	Build user confidence and increase use of digital networks and applications	• Enhancing consumer protection • Discouraging cyber crime • Ensuring data and privacy protection • Enforcing jurisdiction • Developing dispute resolution • Improving information network security • Protecting intellectual property rights

Source: Guermazi and Satola 2005.

- infrastructure
- applications
- consumer confidence (table 2.2).

Developing infrastructure requires public and private investments, investment incentives, and investor protection mechanisms. Providing affordable access to ICT services requires a competitive market with multiple operators and service providers. Strong regulation, government-managed convergence, and equitable distribution of spectrum licenses by the government can help expand and modernize ICT networks.

Providing government services and conducting business online require adapting legal frameworks to an environment where transactions are conducted over electronic platforms. E-transactions raise a number of legal issues unique to the online world, ranging from acceptance of e-signatures to contract formation to the admissibility of electronic evidence to jurisdiction. Intellectual property and copyright laws are important not only to encourage people to invest and work in R&D, but also to assure foreign companies that bringing new technologies and product development processes will not result in loss of competitive advantage.

Finally, deploying ICT applications requires institutional capacity to support network security and an environment of trust where consumers feel safe conducting business online. A secure online environment requires a host of legal supports, such as for data protection and privacy, and consumer protection.

Recent Advances in ICT Laws and Regulations

China has made some progress in adopting laws and regulations addressing each of these three areas.

Infrastructure: Telecommunications Regulations

In 2000 the government outlined new Telecommunications Regulations as an interim step toward modernizing the regulatory regime. These mainly refer to the market (licensing, interconnection, and tariffs), services, construction (facilities and network access), and security. The regulations aim to introduce competition, fairness, transparency, and nondiscrimination as principles for China's telecommunications market. The 81-article regulations define basic and value-added services. Licenses for all services that cover more than two provinces will be issued by the State Council. The responsibilities, including universal access obligations, are detailed in the regulation for service providers. Restricting users' ability to select other providers, using unreasonable cross-subsidies, or charging below-cost prices to drive out competition are considered illegal.

Applications: Regulation of Internet Content

In 2000 the Chinese government issued a regulation on Internet content explaining the difference between general and specific information as applied to Internet news, content, movies, television, and entertainment. General information conveyed over the Internet is expected to be consistent with state and constitutional statutes, preserve law and order, and uphold social harmony. Specific information, however, must abide by regulations stipulated for Chinese media, publishing, information, and broadcasting sectors.

In 2004 the government passed the E-Signature Law, which took effect in April 2005 and granted legal validity and therefore enforceability of a variety of online transactions, providing a secure legal foundation for China's e-commerce industry and encouraging ICT development (box 2.1). This happened in the context that as international e-commerce activities increase, harmonized and interoperable legal and regulatory frameworks will become more necessary. In East Asia and the Pacific, for instance, the Asia-Pacific Economic Cooperation Blueprint for Action on Electronic Commerce was adopted in 1998, and the e-ASEAN (Association of South East Asian Nations) Initiative was endorsed in 1999. They have had positive effects, encouraging the enactment of such laws in member countries (table 2.3).

E-commerce has also been supported by the enactment, in 2000, of interim regulations allowing securities firms to engage in online transactions. China's stock exchanges—such as Shenzhen's Securities Exchange and Shanghai's Securities Exchange—now conduct transactions using IT systems. In 2003, phone and Internet transactions resulted in stock trading of $120 billion, or nearly 15 percent of the

Establishing an Enabling Legal and Regulatory Environment

> **Box 2.1 China's E-Signature Law**
>
> To enable e-commerce transactions and instill confidence in them, the Standing Committee of the National People's Congress enacted the Electronic Signature Law in August 2004. The law was the first national law on e-commerce (Guangdong province had passed such a law in 2002) and became effective throughout China in April 2005. The law provides definitions and guidelines for standardizing the conduct of e-signatures, confirming their legal validity, and safeguarding the legal interests of the parties involved in certain civil and commercial transactions conducted online. The law also addresses issues specific to electronic contracting that were not covered by other legislation. The E-Signature Law grants legal validity to "electronic documents" (data messages as used in the law) and "electronic signatures."
>
> Among its main features, the law also provides requirements for the authentication, dispatch and receipt, and enforceability of electronic documents, the legal effect of e-signatures (and, where necessary, certification of the reliability of e-signatures), and liability for noncompliance. For example, the law provides that an electronic signatory shall be held liable for damages to a party that relies on such a signature or to a supplier of electronic certification, unless the signatory proves that it was not at fault.
>
> Although in its nascent stages and still requiring implementation with respect to certification processes, China's E-signature Law is an important step toward standardizing these issues on a national basis. It offers a general framework, providing flexibility for new procedures and provisions to be added, while giving validity to electronic transactions.
>
> *Source:* Letner 2005.

Table 2.3 E-Commerce and E-Signature Legislation in East Asia and the Pacific

Country	Law on e-commerce and e-signature	Year
Cambodia	Draft Subdecree on Electronic Transactions	
China	Electronic Signature Law	2005
Indonesia	Bill on Electronic Information and Transactions	
Korea, Rep. of	Framework Act on Electronic Commerce	1999
	Digital Signature Act	1999
Malaysia	Draft in legislative process	1997
Mongolia	Draft Law on Information Technology	
Myanmar	Electronic Transactions Law	2004
Philippines	Electronic Commerce Act	2000
Thailand	Electronic Transactions Act	2001
Vanuatu	Electronic Transactions Act	2000
Vietnam	Electronic Transactions Act	2005

Sources: Satola; UNESCAP 2004, Sreenivasan, and Pavalasova 2004.

Note: Bills and laws with no year indicated are at a draft stage or under deliberation.

total. Provincial regulations have also encouraged e-commerce, including those issued in 2004 by the government of Guangdong.

In the international context, China will be faced with a decision on whether to accede to the Convention on Use of Electronic Communications in Electronic Contracts (E-Contracting Convention) recently promulgated by UNCITRAL (United Nations Commission on International Trade Law). The convention covers matters of cross-border e-commerce in much the same way that China's national E-Signature Law covers certain aspects of its domestic e-commerce activities. The purpose of the convention is to remove obstacles to cross-border e-commerce arising from different national rules, as well as reconcile a number of international trade agreements. The convention creates a set of default rules that enables electronic transactions rather than adopting a regulatory approach. The convention grants legal recognition to contracts formed online and to the use of electronic records and signatures, recognizes automated contracts, and provides default rules for dispatch and receipt of electronic communications and for human error. It preserves freedom of contract (Satola 2006).

Resolution on Internet Security

The Internet information services regulations establish criteria and standards for permitted use of the Internet and make service providers liable for the content they display. The computer information network and Internet security, protection, and management regulations require nontransferable registration of all users.

Under the People's Congress resolution on Internet security passed in 2000, the following actions are considered criminal: illegally intruding into state, national defense, or scientific establishments' information systems; intentionally producing or disseminating programs—such as computer viruses—that attack computer systems and communication networks; and interrupting service network or system services. In addition, sales of antivirus and firewall software and spyware and adware detection software require licenses.

Information security products based on encryption technologies have become widely used for commercial information systems. Under the regulation of coding for commercial use, only designated organizations can conduct research and development (R&D) and manufacturing of commercial encryption. Sales of such products must occur under licensing arrangements, and individuals and organizations can only use commercial encryption products approved by the state.

Selected ICT laws and regulations issued in the past 10 years in China are summarized in table 2.4.

Movement Forward

Despite progress in developing the enabling environment, as evidenced by adoption of supporting legislation, China's legal and regulatory framework for informatization faces several challenges. The legal framework aspects of the enabling environment are

Establishing an Enabling Legal and Regulatory Environment

Table 2.4 Selected ICT Laws and Regulations in China

Year enacted	Name	Overseeing entity	Scope
1997	Criminal Law	Courts	Unauthorized access to computer systems
1997	Computer Information Network and Internet Security, Protection, and Management Regulations	Ministry of Public Safety	Regulation of Internet content
1997	Interim Regulation of Computer Networks Connection to the Internet	MII	Adoption of Internet Protocol
1997	Regulation of Broadcasting and TV Administration	SARFT	Regulation of standard TV and radio stations and their relay stations, satellite TV and radio networks and their upload and relay stations, microwave stations, and cable TV and radio stations
1999	Contract Law	Courts	E-commerce: data messages defined as written form
1999	Regulation of Coding for Commercial Use	Office of the State Commercial Cryptography Administration	Specification that only designated organizations can conduct R&D and manufacturing of commercial coding
2000	Customs Law Amendment	General Administration of Customs	E-commerce: allowance of electronic customs declarations
2000	Interim Regulation of Online Securities Trust	Securities Regulatory Commission	E-commerce: allowance of online securities exchange
2000	Regulation of Internet Information Services	MII	Definition of "information" as general or specific
2000	Resolution on Internet Security	Courts	Definition of online crime
2000	Telecom Regulations	MII	Regulation of network infrastructure and telephone services
2002	Interim Regulation of Internet Publishing Management	Press and Publication Administration	Regulation of published content
2004	Regulation of Online Audio and Video Programs	SARFT	Regulation of online audio and video news, entertainment, and TV and radio programs
2005	Electronic Signature Law	MII	Legal validation of data message and regulation of certification authentication
2005	Regulation of News Information Service of Internet Provider	State Council Informatization Office	Regulation of online text news

Source: Author.

quite complex (contributing to the institutional and regulatory competition and overlap) as well as, in some cases, underdeveloped.

Several areas of the legal framework will require high-level legislation in the short and medium term. The People's Congress and its Standing Commission have passed only two major statutes in this area: the E-Signature Law and the Resolution on Internet Security. Other statutes provide only a broad foundation for administrative regulation and departmental regulatory agencies. In some cases the government has been taking a bottom-up approach to legislation—allowing precedence and practices to take hold first, then legislating practices. This approach is useful for legal reform, since it provides rich experiences based on current legal practices and ensures a smooth legislative transition. However, it carries greater risks because it is not systematic and often creates conflicts among departmental jurisdictions. In other areas, a top-down approach has been taken.

China's ministries and regulators involved in ICT development have discussed developing an overarching law to provide a framework for laws and rules on informatization. For several reasons, consideration of such a law remains premature. However, if such a framework were to be considered, it may not be premature to begin formulating policy at an appropriate level to establish the foundation for the consideration of an integrated, comprehensive national legal framework. Many applications are associated with informatization, and their scope is ever-changing. New issues and challenges are constantly surfacing, and better understanding is needed of what legal and regulatory reform is required. In addition, while certain areas of the development of informatization in China may require a "regulatory" approach, China could benefit from international experience and consider, where appropriate, either adoption of industry norms or, as has been done in the UNCITRAL E-Contracting Convention, establish default rules that will apply instead of establishing more regulation.

Noting Major Areas of Reform

Areas where legal and regulatory reform can support informatization in China include telecommunications, network security, Internet content regulation, data protection and privacy, open access to government information, strengthening of protection of intellectual property rights, and cyber crime.

- *Telecommunications.* A new telecommunications law is in the legislative process, but it has not yet been reviewed by the National People's Congress. Regulation of market entry, interconnection, pricing, and spectrum allocation functioned well five years ago but requires revision. China's telecommunications sector has experienced rapid technological advancement and growth in new services. Market entry requirements need to be reconsidered to support further competition and innovation. In addition, the market now involves a host of new players, creating interconnection issues that should be addressed by the legal system.

Establishing an Enabling Legal and Regulatory Environment

With increasing competition and more foreign service providers entering the domestic market, new legal and regulatory issues are expected to emerge. Finally, the provision and framework of universal service obligations need to be addressed.

Many countries are starting to combine regulation of telecommunications networks with information technology and broadcasting. The European Union has largely taken this approach, enabling consolidation of existing principles and integrated execution of ICT laws—as evidenced in the 2002 directives comprising the new *acquis communautaire*. One of the key features of the new *acquis* in Europe is the de-emphasizing of sector-specific regulation and, in recognition of the convergence of all electronic communications, more reliance on competition principles in the broader, more complex electronic communications market.

In 2003 the United Kingdom combined regulatory functions under the Office of Communications, to oversee converging industries and manage the complex dynamics of competition in both content and the communication networks that carry services (box 2.2). This approach resolved issues associated with multiple administrative units' execution of the same laws.

Box 2.2   The United Kingdom's Converged Regulator

The United Kingdom's converged regulator, the Office of Communications (OFCOM), became operational in late 2003. OFCOM brought together several agencies: the Office of Telecommunications, Broadcasting Standards Commission, Independent Television Commission, Radio Agency, and Radiocommunications Agency. This consolidation was designed to streamline regulation of converging communication technologies.

The new regulatory framework consists of the following:

- the existing Competition Act, governing matters related to anticompetitive activity, and the monopoly provisions of the Fair Trading Act, applied jointly by the Office of Fair Trading and the new regulator

- a new, independent statutory regulatory body (OFCOM) responsible for economic regulation of communications, content regulation, and spectrum management

- an open, participatory approach supported by research and a consumer panel to advise OFCOM, enhanced by mechanisms such as citizen juries to address content issues

- co-regulatory or self-regulatory initiatives, developed with OFCOM to deal with issues (such as offensive content on the Internet) where such approaches, backed up as necessary by statutory powers, offer the best means of achieving regulatory objectives

Source: http://www.communicationswhitepaper.gov.uk; http://www.ofcom.org.uk.

Similarly, in China a new law for telecommunications (perhaps more broadly, electronic communications) could also support the development of a level playing field and a fair, open market. In many countries' telecommunications markets, the number of monopolies has fallen, and incumbents have been forced to compete with a range of service providers. Competition has also led to the development of value-added services. Regulators of competitive markets are focusing on a few priorities: expanding value-added services while maintaining pricing policy and interconnection quality for basic services; integrating voice, data, and video services; encouraging network integration; and pursuing universal service coverage. Such priorities should guide the development of China's telecommunications market and laws.

- *Network security.* Information network security law focuses on ensuring the stability of China's computer networks, which are under constant threat from viruses and other menaces, as well as from hackers trying to access government and private networks and data. Accordingly, China needs stronger policies, laws, and consultative processes to ensure the security of networks (World Bank 2006b).

- *Internet content.* China's Internet content regulators determine whether content is consistent with social values and focus on preventing dissemination of information that the government considers illegal. Unlike information network security, which mainly deals with data and network protection, Internet content regulation involves placing regulations on what Chinese people read, see, and hear on the Internet. While regulation occurs at different levels and the existing legal framework results largely in self-censorship, advances in technology effectively make content regulation on the Internet an increasingly costly enterprise for the government. Movies, television, and newspapers and other print media—which have central points of distribution—are easier to regulate than Internet content, which is created and distributed by multiple sources. In addition, print media can often be found online, requiring that agencies regulate content both online and offline.

- *Data protection and privacy.* Abuse of personal information has become an issue in China's developed urban areas, yet there is no specific law protecting consumers. Releasing personal information that does not infringe on the reputation of the subject is not considered illegal. To avoid harassment, consumers often provide false information or refuse to give personal information online—hampering the development of China's credit reporting system.

- *Open access to government information.* The Chinese government is interested in providing the public with more information and considers increased transparency and accountability crucial to eliminating corruption. Still, despite efforts to introduce freedom of information programs, China does not have a specific law requiring the government to provide information.[1] Moreover, the cost of accessing information is still relatively high. Although the government has built its capacity to provide information to the public, further legal reform is needed to make information administration more open and effective. A government

Establishing an Enabling Legal and Regulatory Environment

information access law would provide a legal basis to support the free flow of information and the development of e-government.

- *Intellectual property rights.* One of the principal remaining causes for concern in the area of protection of intellectual property rights (IPRs) is piracy, which poses an obstacle to conducting business in China for both local and foreign investors. Even though China relies heavily on open-source software (for example, to circumvent piracy issues related to proprietary software and reduce dependence on foreign proprietary code), the essence of the "openness" of open-source software relies entirely on robust copyright protection. Accordingly, China is now realizing that its own interests are best served by addressing these issues. The government has recently taken steps in the computing and software areas, adopting directives to safeguard software copyright. The next challenge will be to transpose these into actual law.

- *Cyber crime.* The main issues to be addressed in specific legislation dealing with cyber crime are illegal data interception, data interference, and system interference and access, as well as other crimes facilitated by computers. The Criminal Code already makes unauthorized access to computer systems illegal. It should also be clarified that other criminal matters in the physical world (forgery, fraud, copyright infringement, and production and distribution of obscene or pornographic material) apply equally to the virtual world of the Internet as they do to the physical world. Some of these may be dealt with through amendments to existing laws, and others may require new laws.

Criminal law could be similarly amended to include cyber crime, and provisions on acknowledgment, protection, and management of IPRs in cyberspace could be added to general IPR laws (Patent Law of 1984, Trademark Law of 1984, Copyright Law of 1990, and Contract Law of 1999).[2]

These seven areas—telecommunications, network security, Internet content, data protection and privacy, open access to government information, intellectual property rights, and cyber crime—and implementation of the recent E-Signature Law are the priorities for the informatization legal system in China. Additionally, the enabling environment also needs to be improved by amending existing laws.

Improving Certainty and Coordination of Regulatory Responsibilities

A second challenge is to address the complexity and resulting uncertainty of the division of regulatory responsibilities and their lack of coordination. The absence of a legal framework stipulating the principles and scope of informatization has made regulations unclear. Cooperation among regulatory departments is weak, and China's myriad agencies have different and sometimes overlapping responsibilities. As a result, many agencies often regulate the same area—yet accountability is lacking, and the resulting regulations are inconsistent. For example, the definition of "illegal content" was mentioned in regulation of Internet information services, regulation to protect computer system security, (interim) regulation of Internet publishing, and regulation of online audio and video, overseen by different entities. This has made the implementation and execution of laws and regulations problematic. The legal system should strengthen the rule of law to ensure that laws and regulations are enforced and enterprises and government agencies are held accountable.

Balancing Rights and Obligations

Finally, the legal system structure is unbalanced in terms of rights and obligations. Many existing laws and regulations are designed from the perspective of administrative convenience, often considered more important than individual or enterprise rights. Among enacted laws and regulations, there are far more restrictive provisions than other measures, such as for self-regulation and dispute resolution. Obligations are given more weight than the legal rights of subjects. An emphasis on processes and procedures (such as licensing) has overlooked protection of privacy and individual rights. Inadequate attention is paid to personal data protection and copyright protection. An enabling legal and regulatory system should focus on encouraging innovation and avoid restrictive provisions on R&D that hinder technological change and economic development.

One remedy would be for the government to promote consultation and participation by citizens and business units in its legal and policy-making processes. The government has developed several ways to encourage cooperation with business and civil society on legal matters (box 2.3), but there is considerable room for improvement. Due to the lack of a freedom of information act, the private sector and civil

---

Box 2.3 Legislative Mechanisms in China

China's current legislative mechanisms are as follows:

- *Legislative resolutions.* China's legislature comprises the People's Congress and its Standing Commission. With the signatures of 30 delegates of the Congress, a legislative resolution can be presented to the Congress for discussion. For example, 32 delegates presented a resolution on e-commerce in 2000.

- *Legislative drafting and research.* Chinese laws can be drafted in several ways. The People's Congress can authorize a research institute, academic institute, government agency, or nongovernmental organization (NGO) to draft a proposal. During drafting, the organization responsible can hold public seminars and hearings; collect ideas and opinions from private entities; and request comments from government departments, enterprises, and industrial associations.

- *Legislative review.* When the People's Congress and its Standing Commission review a draft law, delegates can offer comments and ask questions that must be further researched. For example, during the review of the E-Signature Law, members of the People's Congress conducted surveys in Beijing, Shanghai, and Guangdong, and representatives of enterprises and industry associations were invited to provide comments.

Establishing an Enabling Legal and Regulatory Environment

society are unaware of government deliberations on policy. Without public participation, it is difficult to build a sense of legitimacy and representation of proper rights. Thus an open approach should be used to formulate informatization policies. Furthermore, e-government applications can be used to increase public participation in policy making.

To help the country to realize its informatization strategy, China must provide certainty and predictability in its legal and regulatory system by fostering a business-friendly climate where competition can flourish, foreign and domestic sponsors can invest with confidence, and ICT development can take root. As China continues to build and improve its ICT enabling environment, obstacles to economic growth and development will be removed. These measures ultimately benefiting China should also create conditions for resolving issues relating to informatization, such as issuance of third-generation network licenses, potential restructuring of state operators, development of local technology standards, opening of markets to foreign players, development of systems for universal service obligations, and control and censorship of content and applications over telecommunications networks.

Notes

1. Recently, the State Council has approved the Directive on Government Information Openness.
2. Based on information from the State Intellectual Property Office, http://www.sipo.gov.cn/sipo/.

Chapter 3

Enhancing Telecommunications Infrastructure

China has long tried to promote ICT development by improving its telecommunications infrastructure. This strategy was recently expanded to include the Internet. Between 1995 and 2004, China's gross domestic product (GDP) grew by an average of 8.6 percent a year, reaching $1.65 trillion. The country's telecommunications sector grew even faster (figure 3.1).

Market Overview

As in many countries, telecommunications networks in China were state-owned, with services provided by the former Ministry of Posts and Telecommunications through the Directorate General of Telecommunications and provincial semi-autonomous administrations. Since the 1990s China's telecommunications market has become more liberalized. In 1994 China Unicom was created as a state-owned operator to compete in the long-distance, Internet, and mobile markets. In 1997 China Mobile was spun off from China Telecom, the previous incumbent monopoly, and partially privatized. In 2000 China Unicom was also partly privatized through a listing of its shares on the Hong Kong Stock Exchange. This restructuring paved the way for the explosive growth of the mobile sector because it facilitated competition between China Unicom and China Mobile in mobile services.

Further liberalization occurred in 2002, when the government split China Telecom into two competing operators: China Telecom (serving the south) and China Netcom, a merger of China Telecom (serving the north), China Netcom, and the Jitong Corporation. Basic telecom licenses were issued to five state-owned operators—China Telecom, China Netcom, China Unicom, China Mobile, and China Railcom—with each focusing on different market segments (table 3.1).

The Chinese government's unique entry policy—only network-based competition was introduced—explained the limited entry observed in the telecommunications

Figure 3.1 Telecommunications Investment and Revenue in China, 1990–2004

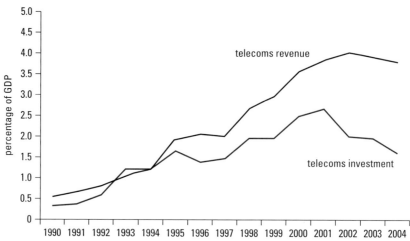

Sources: ITU 2006; World Bank 2006c.

market in China. The goal was to balance the concerns of overinvestment and effective competition, and to nurture four or five large state-owned operators before China's telecommunications market was fully opened to foreign investors, in compliance with China's World Trade Organization commitments (table 3.2). Since then, foreign operators such as Telefónica (Spain) and Vodafone (United Kingdom) have had some minority ownership in the sector.

Among the five major telecommunication operators, China Telecom is largest in terms of employees: in 2004 it had about 300,000, almost the sum of all the other operators. China Mobile is the largest operator as measured by revenues, with turnover of nearly $30 billion in 2005 (figure 3.2). In 2005 the total revenue of these operators was $71.4 billion, about 4 percent of Chinese GDP.

Fixed-Line Operators

China Telecom and China Netcom, the two main fixed-line providers, have 86 percent of the local access market in China. China Netcom provides fixed line services in 10 northern provinces, while China Telecom serves 21 in the south. The other players in the fixed line market are China Railcom and China Unicom, but their market shares are small (figure 3.3a). China Telecom and China Netcom have slowly tried to enter one another's territory, especially with Voice over Internet Protocol (VoIP) and data services. China Netcom has begun to target customers outside its original northern service area and now offers services to residential and corporate customers in Shanghai and Guangdong province. Similarly, by late 2004 China Telecom claimed a 5.5 percent market share in China Netcom's northern territory (BMI 2005).

Enhancing Telecommunications Infrastructure

Table 3.1 China's Main Telecommunications Providers, by Market Segment

		Market segment					
		Fixed			Mobile	Data	Internet
Provider	Ownership	Local	DLD	ILD			
China Telecom	China Telecommunications Corporation (71 percent), public (17 percent), others (12 percent)	X (southern China)	X	X	Limited mobility services	X	X
China Netcom	China Netcom Holdings (70.5 percent), public (20 percent), Telefónica (5 percent), five PRC shareholders (4.5 percent)[a]	X (northern China)	X	X	Limited mobility services	X	X
China Mobile[b]	China Mobile (76 percent), Vodafone (3.3 percent), public (20.7 percent)				X		X
China Unicom[b]	China Unicom (74.6 percent), public (25.4 percent)	X	X	X	X	X	X
China Railcom	Ministry of Railway (51 percent), 15 subsidiaries (49 percent)	X	X			X	X

Sources: BMI 2005; China Netcom 2005; China Telecom 2005.

Note: DLD = domestic long distance; ILD = international long distance; PRC = People's Republic of China

a. The five shareholders are the Academy of Sciences; Information and Network Center of State Administration of Radio, Film and Television; China Railways Telecommunications Center; Shanghai Alliance; and Shangdong Provincial State-owned Assets Supervision and Administration Commission.

b. It provides nationwide coverage through 31 regional subsidiaries.

Mobile Phone Operators

The cellular mobile market segment remains a duopoly (figure 3.3b), with China Mobile and China Unicom offering nationwide service. China Mobile operates a global system for mobile communications (GSM) network that had 247 million subscribers at the end of 2005.[1] China Unicom operates two mobile networks (GSM with 95 million subscribers and code division multiple access (CDMA) with 33 million subscribers).

Full competition has not been achieved in the mobile market because of the limited number of competitors. Furthermore, demand-side network effects and supply-side economies of scale mean that China Mobile has significantly more market power.

China's Information Revolution

Table 3.2 China's Telecommunications Commitments to the World Trade Organization

Sector	Phase	Permitted percentage of foreign investment	Date	Geographic limit
Value-added and paging	I	30	2001	Beijing, Shanghai, Guangzhou
	II	49	2002	Extended to 14 other cities
	III	50	2003	Nationwide
Mobile	I	25	2001	Beijing, Shanghai, Guangzhou
	II	35	2002	Extended to 14 other cities
	III	49	2004	17 cities
	IV	49	2006	Nationwide
Fixed line	I	25	2004	Beijing, Shanghai, Guangzhou
	II	35	2006	Extended to 14 other cities
	III	49	2007	Nationwide

Sources: China Mobile 2005; Qiang 2001.

Figure 3.2 Employees and Revenues of China's Main Telecommunications Providers

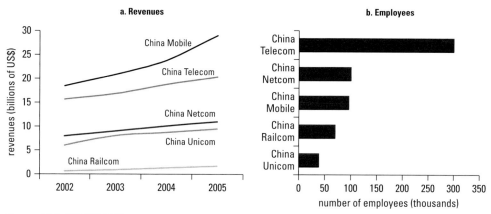

Sources: BMI 2005; MII 2005.

This situation is reinforced by the fact that China Unicom operates two technologically different networks, CDMA and GSM.[2]

Both mobile operators face growing competition from the personal handy phone system (PHS) services offered by China Telecom and China Netcom.[3] PHS allows mobile access to telecommunications services in urban areas. Because they provide limited mobility, PHS services are relatively inexpensive to deploy. As a result, PHS is competitive with mobile services at the low end of the market. With lower and simpler tariffs and calling-party payment schemes,[4] PHS services are at least 50 percent cheaper than competing mobile services. This makes PHS attractive to subscribers who just want a simple, limited mobility service without roaming and fewer features than conventional cellular networks.

Enhancing Telecommunications Infrastructure

Figure 3.3 Market Shares of Fixed and Mobile Providers, 2005

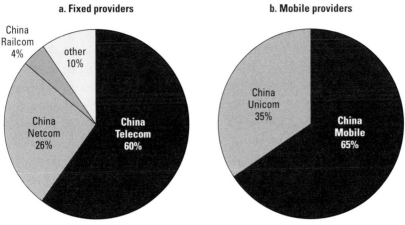

Sources: BMI 2005; IFC 2005.

In recent years, there has been explosive growth in PHS, with China Netcom and China Telecom expanding service in smaller cities across China. As growth in the fixed line market has slowed and consumers increasingly replace fixed with mobile phones, PHS services have become important sources of revenue for fixed line operators.

The future of PHS is unclear given that the fixed line operators do not have mobile licenses. However, given the large number of PHS subscribers—84 million at the end of 2005—the service is emerging as a significant alternative for mobile services. China Telecom and China Netcom have begun offering value-added services such as text messaging, handsets with color displays, and personalized ring tones to their PHS subscribers, making the system more attractive. Some countries such as India and Nigeria eventually allowed such limited mobility services to become full-fledged competitors by granting operators the right to provide nationwide roaming. Should the Chinese government grant additional mobile licenses, it seems certain that the fixed line operators would be interested.

Internet Operators

All five major telecommunications operators operate their own Internet service providers (ISPs). ChinaNet, operated by China Telecom, is the country's largest ISP, providing Internet access and hosting 70 percent of Web sites in China. Local ISPs lease lines from fixed line operators. Prices for ISPs, which vary depending on the quality of service, have been continuously falling because of intense competition.

Although most subscribers use dial-up access to the Internet, broadband access is growing rapidly. About 70 percent of China's broadband users use asymmetric digital subscriber lines (ADSL). China Telecom and China Netcom are the largest operators in the broadband market. In addition to ADSL, they and other providers—Great Wall Broadband, and regional companies—offer services based on local area networks.

Figure 3.4 Telecommunications Penetration in China, 1994–2004

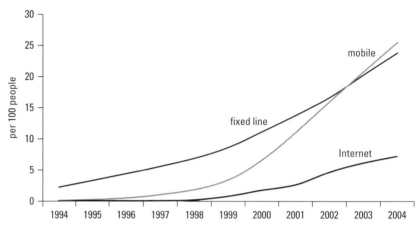

Sources: ITU 2006; World Bank 2006c.

Sector Performance

A decade of rapid development has led to significant progress in China's telecommunications infrastructure (figure 3.4). The country has the world's largest fixed and mobile telephone markets and the second largest Internet and broadband markets, in terms of subscribers.

China's Internet market is migrating from narrowband to broadband access, with the number of broadband subscribers reaching 64.3 million by the end of 2005. Internet content has also developed rapidly, with nearly 50 million computer hosts and 694,000 Web sites in 2005 (CNNIC 2006). Broadband has grown because of falling prices for equipment and subscriptions. Since 2000, the cost of ADSL equipment has dropped from $360 to $35 per line, reflecting growth in domestic manufacturing. In addition, operators have eliminated installation charges, and subscription fees are as low as $3 for 20 hours per month in Beijing (figure 3.5).

Still, Internet access prices remain high as a percentage of income. In 2003 the price basket for Internet use was less than 1 percent of income in developed countries, while it was more than 10 percent in China—higher than the regional average (figure 3.6).

Overall telephone penetration rates—the percentage of the population subscribing to fixed and mobile telephone services—in China remain low relative to Brazil, Russia, and other Asian economies (table 3.3). With 650 million fixed and mobile telephone users in 2004, the national penetration rate was just 50 percent, while in high-income countries it was 133 percent.

Moreover, despite the rapid telecommunications infrastructure development at the national level, substantial regional differences remain. In 2003 the fixed-line penetration rate was 32 percent in the eastern region while mobile penetration was 34 percent (figure 3.7a). However, fixed-line penetration was just 17 percent in central regions and 15 percent in western regions, and mobile penetration was 15 percent in both. Internet development has also been uneven, with nearly 80 percent

Enhancing Telecommunications Infrastructure

Figure 3.5 Charges for Broadband (ADSL) Access in Beijing, 2001–03

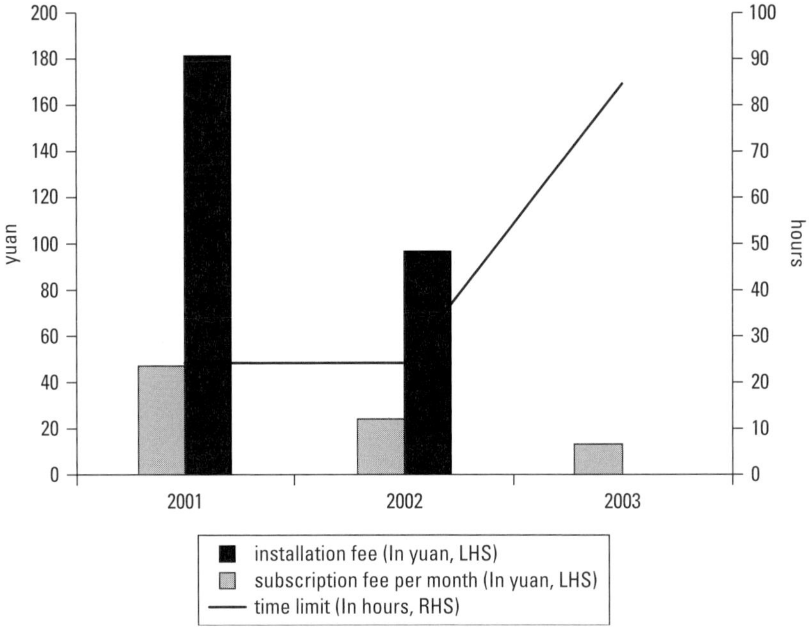

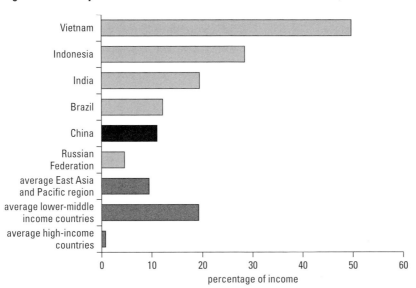

■ installation fee (In yuan, LHS)
▨ subscription fee per month (In yuan, LHS)
— time limit (In hours, RHS)

Source: MII 2004.
Note: LHS = left-hand side; RHS = right-hand side.

Figure 3.6 Monthly Price Basket for Internet Use in Selected Countries, 2003

Source: World Bank staff analysis based on World Bank (2006c).
Note: Price basket for Internet is calculated on the basis of the cheapest available tariff for accessing the Internet for 20 hours a month (10 hours peak and 10 hours off-peak). The cost for telephone line rental is not included, but the cost for the telephone usage charges is included if applicable.

China's Information Revolution

Table 3.3 Telecommunications Penetration in East Asia and Other Developing Economies, 2004 (percent)

Economy	Fixed line	Mobile	Total
East Asia			
Hong Kong (China)	54.8	110.9	165.7
Indonesia	4.8	13.5	18.3
Malaysia	19	62.7	81.7
Philippines	4.1	38.6	42.7
Taiwan (China)	59.1	94.0	153.1
Thailand	10.4	43.0	53.4
Vietnam	6.8	5.7	12.5
BRIC countries			
Brazil	23.7	36.7	60.4
Russia	27.0	50.5	77.5
India	4.2	4.5	8.7
China	**24.1**	**25.8**	**49.9**
High-income average	**55.8**	**76.7**	**132.5**

Sources: BMI 2005; World Bank 2006c.

of the country's domain names registered in northern, eastern, and southern provinces, and the rest in the northeast, northwest, and southwest. Moreover, 10 percent of administrative villages in China, containing 124 million people, had no access to telecommunications in 2003. Three-quarters of these villages were in western regions (figure 3.7b).

In 2003 average disposable income per capita in China's urban areas was about $1,150, while in rural areas it was only $315. Because penetration rates and income

Figure 3.7 Fixed Line and Mobile Telephone Penetration by Region, 2003

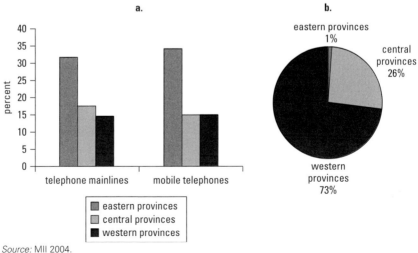

Source: MII 2004.

Enhancing Telecommunications Infrastructure

Figure 3.8 Fixed Line and Internet Penetration in China's Urban and Rural Areas, 2003

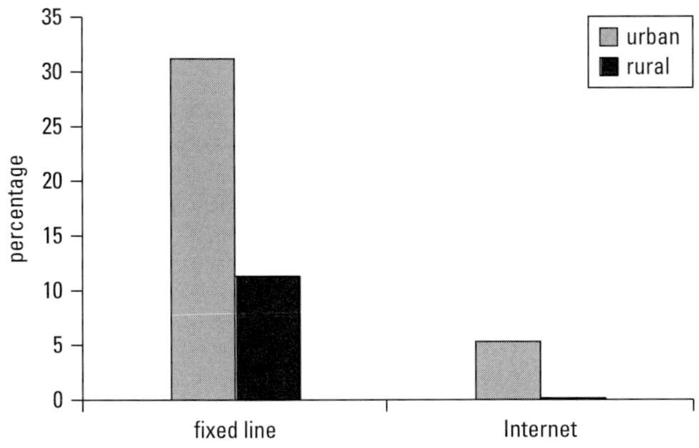

Source: MII 2004.

levels are highly correlated, bridging the digital divide between urban and rural areas (particularly remote areas) will remain a challenge. In 2003, the fixed-line penetration rate was 3 times higher in urban areas; Internet penetration was nearly 40 times higher (figure 3.8).

Strategic Focus

An advanced and widely accessible broadband infrastructure is a prerequisite for successful informatization in China. Many of China's informatization projects in areas such as software development, digital media, e-commerce, and e-government are dependent on a broadband platform to achieve high levels of functionality and usage. The broadband infrastructure must be widely accessible to minimize the digital divide and benefit social and economic development through the availability of rich multimedia applications in fields such as education, health, commerce, and public administration. Broadband networks support the provision of voice, Internet, and broadcasting services over different platforms (fixed line, wireless, and cable television), increasing intermodal competition and resulting in lower prices for users.

Different approaches have been put forth for achieving the twin goals of upgrading to an advanced infrastructure and increasing access. On the one hand, there have been signs favoring the encouragement of domestic resources for infrastructure development through the promotion of national standards. For example, the government has voiced support for the use of national standards in wireless technology areas such as third generation (3G) and wireless local area network (WLAN). In terms of service provision, predominantly state-owned operators have been fortified through limited competition and restricted foreign investment in the areas of fixed line and mobile networks. While these policies are meant to help nurture domestic capability and develop strong national operators, in the long run they could hamper efficiency, investment, and innovation in China's telecommunications service industry. On the other hand, larger participation by foreign companies in China's huge telecommunications market can help attract investment and spur greater competition. It would also make domestic companies more efficient and innovative and hence better able to achieve their own globalization ambitions.

China needs to focus on several key areas over the coming decade to achieve the twin goals of increased access to information and communication technology (ICT) and implementation of state-of-the-art telecommunications infrastructure.

Providing Universal Access in Rural and Remote Areas

When China Telecom was a monopoly, it was required to extend services to high-cost and low-access areas and was allowed to charge more for certain services (such as international calls) to cross-subsidize such expansion. The opening of the telecommunications market and the introduction of competition have increased the number of operators, and network infrastructure development is now based on market mechanisms and profit determinations. Cross-subsidies are hard to sustain in this environment, and operators are finding it costly and difficult to serve rural and remote markets.

Expanding rural ICT access is critical to equitable social and economic development. The Chinese government has paid special attention to rural areas in its 11th development plan, particularly to reduce poverty, inequality, and social exclusion. Hence, China's telecommunications strategy must more vigorously address the gaps between urban and rural areas and across provinces. The telecommunications regulation enacted in 2000 did not set targets for expanding rural access, such as universal service obligations. The regulation, however, does define a general scheme to subsidize operators expanding services in areas where it is unprofitable to do so.

China has not yet established a universal service policy, such as a fund for financing infrastructure expansion in under- and unserved areas. Nor has it started collecting fees from operators to support infrastructure coverage obligations. However, it is exploring other mechanisms for expanding rural access. In 2004 the Ministry of Information Industry (MII) initiated a project to make telephone service available to all villages, requiring the five main telecommunications providers and China Satellite Communications Corporation to share the obligations of universal access based on geographic divisions. Given the importance of the Internet, the government set a goal of providing Internet access to all villages by 2010.

Subsidies to maintain lower service charges could increase rural penetration. Alternatively, competitive bidding among operators for subsidized contracts would help define obligations and deliver services transparently.

Other methods can also be used to increase rural subscribers. Phone bills could be sent more frequently, enabling subscribers to pay smaller amounts on an incremental basis. In addition, charges for outgoing calls can be prepaid. After the balance has been used up, outgoing service (except for emergency numbers) would be denied but incoming calls still allowed. This approach prevents subscribers from running up

Enhancing Telecommunications Infrastructure

large telephone bills, being unable to pay them, and falling into debt—a fear among those with low incomes. This is a method typically used for mobile prepaid cards where subscribers can continue to receive calls for a while even though the card value has been used up. Indeed, there is significant scope for encouraging mobile operators to enhance coverage and widen access to mobile services in China. While practically all cities and major roads are covered by mobile service, it needs to be extended to the less widely served central and western regions of the country.

In any case, the government should adopt an open policy to allow for technology-neutral ways of implementing universal access obligations.

Developing Broadband Infrastructure

Broadband Internet access is a clear strategic direction for China's telecommunications infrastructure development. Yet several barriers impede further expansion of the broadband market. Many potential users remain uneducated about the service or lack the capacity or income to use it. In addition, broadband access depends on personal computer (PC) ownership. In 2003 about 28 percent of urban Chinese households owned a PC, contrasted with slightly more than 1 percent of rural households. While PC penetration in Beijing reached 23 per 100 people in 2004, 21 provinces were below the national average (figure 3.9). Without increased PC ownership in rural and western regions, demand for broadband will be limited to high-income urban areas. Policies to promote wider PC penetration include payment plans stretched out over a long period, as well as the development of inexpensive, entry-level computers.

Broadband development will remain a top priority of China's telecommunications infrastructure development. The government should take steps to intensify intermodal competition among fixed line providers, cable television companies, and wireless operators to reduce broadband costs, making service affordable to more consumers and enterprises. This approach includes waiving existing restrictions that prohibit operators from providing "triple-play" services (voice, Internet, and broadcasting) over their broadband infrastructure.

Broadband also needs to be extended to rural areas. One challenge is that most villages without phone access are in mountainous and remote regions. The northeastern region is scarcely populated, while the difficult terrain in rural Guizhou, Sichuan, and other southwestern provinces presents severe challenges for cost-effective communications coverage.

Finding appropriate technology to connect such areas is challenging and costly. Satellite technology using very small aperture terminals (VSAT) could potentially be effective for providing services in these areas. VSAT can provide service to large areas where it would be prohibitively expensive to lay fixed lines or install the many cellular towers required to cover scattered populations. For instance, VSAT has been successfully introduced in Tibet and Xinjiang provinces, where almost 3,000 terminals provide telephone access across a wide area of western China. However, economies of scale should be taken into consideration for the deployment of VSAT, because it would work effectively only under a situation of sizable demand in large areas. VSAT's

Figure 3.9 Personal Computer Penetration in Selected Provinces in China, 2003

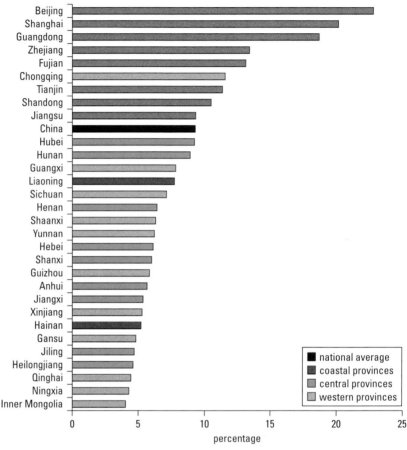

Source: World Bank 2006a.

operating costs are about four times as much as fixed or mobile lines in normal situations, which limits its potential to be adopted in remote areas with sparse demand.

Another alternative is low-frequency mobile technology, which can cover a wider area than the 900 megahertz (MHz) and 1800 MHz systems presently used by China's mobile operators. Technologies based on 450 MHz using CDMA have been deployed in a number of countries to expand service in rural areas. Certain implementations of CDMA-450 can achieve broadband speeds of close to 2 megabits per second (Mbps). China's synchronous code division multiple access (SCDMA) technology offers an alternative to CDMA-450. SCDMA can operate at low frequencies (440 MHz) and provide service over 40 kilometers per tower. SCDMA, however, may take several years before deployment in rural areas. Nevertheless, the MII has recommended using SCDMA to meet universal service obligations, indicating support for its development. The government faces the challenge of promoting a domestic technology that may in the long-run achieve economies of scale and

Enhancing Telecommunications Infrastructure

hence lower investment costs versus accepting technology-neutral solutions that could rapidly promote the development of broadband infrastructure.

Constructing Third-Generation Telecommunications Networks

Third generation (3G)[5] is the generic term for the next generation of wireless mobile communications networks. 3G networks transmit data at broadband speeds, typically up to 2 Mbps.[6] There are three main incompatible standards for 3G recognized by the International Telecommunication Union. In addition to technical differences among the standards, strong political pressure exists. Wideband-code division multiple access (W-CDMA) was proposed by European Union firms and promoted as the successor of GSM systems; CDMA-2000 was promoted by QUALCOMM, the U.S.-based company with many patents in CDMA technology; and time division-synchronous code division multiple access (TD-SCDMA) has been promoted by Chinese company ZTE, with Datang Mobile and the China Academy of Telecommunications Research of the MII.

In 2004 the MII and its Academy of Telecommunications Research completed technical trials for all three standards. No decision has been made about the allocation of China's 3G licenses, a decision bound to be colored by the government's wish to involve the country's homegrown standard, TD-SCDMA,[7] and by continuing uncertainties over the restructuring of the major telecommunications players. China Mobile, which currently operates a GSM network, may receive a license for W-CDMA,[8] and China Unicom is expected to upgrade its network by deploying a new CDMA EV-DO (Evolution Data Only) technology. But it has not been decided whether both fixed line operators (and PHS operators), China Telecom and China Netcom, will receive 3G licenses or which technologies they might deploy.

The adoption of 3G technology would spur enormous network infrastructure development, enable telecommunications and other service providers to offer a wide range of new applications and content services, provide another option for broadband access, and help ensure that China's mobile networks do not fall behind in technical sophistication. However, uncertainties about the technology that the government will select, as well as the timeframe it will establish for rollout, have created uncertainty among equipment manufacturers and service providers.

To promote 3G development, the Chinese government may want to consider a number of options. For instance, so-called mobile virtual network operators could be allowed to enter to increase competition at the service layer. At the same time, competition at the infrastructure layer is important to spur innovation. The choice of the type of 3G system to implement should be technology neutral and left to operators with the government awarding the necessary spectrum. To help offset the high cost of investment in 3G networks, the government could encourage infrastructure sharing, a policy adopted in countries such as Australia and Sweden. A more open entry policy is crucial to mitigate potentially significant risks in 3G network deployment and to promote rapid development. China cannot afford to further delay implementation of 3G networks given the many potential benefits.

Promoting Investment in Telecommunications Networks

Investment in China's telecommunications networks totaled over $25 billion in 2004—the fifth consecutive year with investment exceeding $25 billion. While investment has been sustained, network development has been hardware-intensive, focusing on infrastructure and equipment. Less attention has been paid to operations management and support systems, customer service, and network optimization. Despite global equipment standards, China has not standardized its network technologies. Operators must increasingly focus on these areas.

The government sometimes chooses technologies to promote Chinese standards (e.g. SCDMA) in support of industrial policy. This has a direct impact on the operators' investment decisions and the strategies of equipment manufacturers.

The government also influences technology selection through the issuance of licenses and affects the scale of investments through the review of operator plans. Once 3G licenses are issued, China Telecom estimates that it will cost $9.6 billion to deploy a nationwide 3G network. While supporting its own industry has some merits, the government should rely more on market forces in the selection of adequate technologies, both for 3G deployment and for universal access implementation.

Upgrading telecommunications infrastructure and expanding access will require significant investment over the coming years. It is uncertain that existing domestic operators will have the resources necessary to fund the needed investment. Therefore, the government needs to encourage greater competition and the participation of strategic foreign operators to attract more investment into the sector.

Encouraging Network Convergence

Convergence involves integrating services across different platforms such as telephone, Internet, mobile, and broadcasting. For example, fixed line telephone networks can now offer video services over broadband technologies such as ADSL, and cable television networks can provide Internet access.

China's 11th Five-Year Plan is expected to advocate eliminating further barriers to convergence. This will be a long process, involving convergence between fixed and mobile telephony, the Internet, and broadcasting.

China's telephone operators, content providers, equipment manufacturers, and research institutions have been supporting the integration of fixed line networks and broadband networks—that is, the convergence of voice and data services. Accordingly, fixed line operators are investing in Internet infrastructure to deliver integrated services. The rise of broadband and VoIP raises questions about further investment in standard fixed-line networks.

To support convergence, the Chinese government will need to address policy and regulatory issues that prevent broadcast, broadband, and telecommunications providers from accessing each other's networks. Some government reorganization may also be needed. For instance, the MII and the State Administration of Radio, Film, and Television have some overlapping areas of responsibility (see chapter 2).

Enhancing Telecommunications Infrastructure

Establishing an independent ICT regulator would be a crucial step to support convergence. Since network mergers involve interconnecting various systems, technologies, and standards, uniform approaches are needed. China may also need to revise radio spectrum management policies to support network convergence.

Finally, continuous research and development is needed in technologies that enable network convergence. China's government, research institutes, operators, equipment manufacturers, and content and service providers need to establish a comprehensive, multistakeholder process to determine key convergence goals for developing the necessary hardware and supporting applications.

Notes

1. GSM is a digital cellular phone technology, first standardized in Europe, and which today has the largest number of subscribers in the world. Code division multiple access (CDMA) is a digital cellular phone technology that uses spread-spectrum techniques developed by QUALCOMM.
2. In January 2005, China Unicom announced that it would shift investment focus from CDMA to GSM in 2006; it has in the past invested significantly in its CDMA network, the capabilities of which have exceeded demand (Global Insight 2006).
3. Only a few countries use PHS services, which provide mobility similar to GSM or CDMA. In China, intracity PHS service is allowed, but intercity roaming of PHS is prohibited.
4. China has adopted a non-reciprocal interconnection policy between fixed-line and mobile operators, closely related to the receiving-party-pays principle for the mobile sector. Specifically, PHS is considered local fixed line service, so the fixed line operator does not need to pay a mobile termination fee for a PHS-to-mobile call. However, for a mobile-to-PHS call, the mobile operator needs to pay the fixed line network a termination fee. This asymmetry has not yet been resolved.
5. 3G is defined in this report as broadband mobile (faster than 256 kilobits per second).
6. A new development, High Speed Downlink Packet Access boosts speeds for some 3G systems to a theoretical 14 Mbps, although in practice the highest speeds achieved by 3G networks are presently around 2 Mbps.
7. In late 2005, the TD-SCDMA was approved by China's 3G Partnership Project as one of the country's 3G communications standards.
8. W-CDMA has been the typical 3G technology for GSM networks. By the end of 2005, 80 operators around the world offered W-CDMA mobile services.

Chapter 4

Developing and Innovating the ICT Industry

China's information and communication technology (ICT) industry influences the role of technology in the country's economy and society and affects efforts to build an information society. ICT markets contribute to gross domestic product (GDP), generate export revenues, create jobs, and produce consumer goods that meet domestic demand.

China's government has been promoting the ICT industry since the mid-1980s by supporting domestic companies and research and development (R&D) efforts. In addition, the government has provided incentives to attract foreign investment—while also requiring foreign firms to transfer technology in return for market access. Given the size and potential of China's market, many foreign firms have been willing to make this tradeoff, even under conditions they would not accept in other countries.

The Chinese ICT sector has grown rapidly. The domestic ICT market was valued at $156 billion in 2004, up 13.2 percent over the previous year.[1] According to the Organisation for Economic Co-operation and Development (OECD), China led the world with $180 billion of ICT exports in 2004, surpassing the United States.[2] At the same time, the Chinese ICT market is characterized by several ironies:

- Although China is a leading exporter of semiconductors, it still needs to import chips to meet local needs.

- Although Chinese firms export third-generation (3G) mobile technologies, these same technologies have yet to be commercialized in the mainland.

- Although Lenovo, China's leading personal computer (PC) firm, is the third largest in the world, there are only an estimated 65 million computers in a country with a population of over 1 billion people.

This chapter discusses four strategic ICT markets: hardware, software (including services and outsourcing), information and network security services, and digital media—and the policies needed to develop the industry by stimulating innovation and supporting R&D.

Hardware

China's hardware market was $23.6 billion in 2004. Two of its strategic sectors are integrated circuit (IC) production and PC production.

Integrated Circuit Industry

Sales revenue for the domestic IC industry has grown quickly in recent years, jumping almost 200 percent between 2000 and 2004 to reach $6.7 billion. Such growth reflects the introduction of new products by domestic design houses and the opening of new foundries (where chips are manufactured). Strong domestic demand for semiconductors, government backing (box 4.1), and venture capital have also contributed to the rapid growth of China's IC design market. Today there are over 450 IC design firms in China (Yu 2005). Since about 90 percent of IC demand in China is met by imported chips, testing, packaging, and assembly accounts for the lion's share of industry sales (figure 4.1).

The rapid development of foundries is creating an environment where IC design firms can begin to succeed (box 4.2). Proximity to foundries lowers costs, makes it

---

Box 4.1   Government Initiatives toward the Integrated Circuit Industry

In the 1990s limited investment was identified as the primary reason for the slow growth of China's IC industry. Now the government is inviting foreign investors to fuel the industry's growth. According to Fabless Semiconductor Association, the industry received $3.5 billion in investment between 2000 and 2002—equivalent to the total of all previous investment in the sector.

The government views semiconductors as a strategic industry and has offered extensive support to companies involved in IC design and manufacturing. Support has come from central and provincial governments and science parks.

As for promotional policies for the IC industry, the government published Document 18 (Policies for Encouraging the Development of Software and Integrated Circuit Industries) in 2000 and proposed several actions to support the industry, including publicly subsidized bank loans, government investments, tax breaks, and funding for design centers.

Central government policies are complemented by local incentives to attract IC manufacturers. Local governments often compete to attract investment. For example, Beijing offers a *Shanghai + 1* policy to add a year to any financial incentive offered by Shanghai.

Source: IFC 2005.

Developing and Innovating the ICT Industry

Figure 4.1 Sales Revenues for China's Integrated Circuit Industry by Segment, 2004
($ millions %)

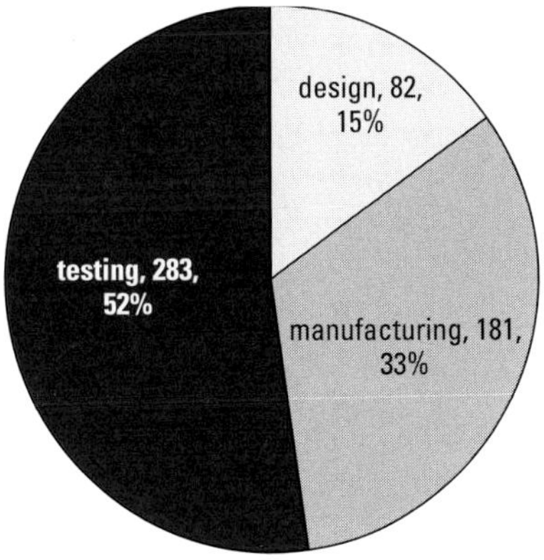

[Pie chart showing: testing, 283, 52%; manufacturing, 181, 33%; design, 82, 15%]

Source: Yu 2005.

easier for design firms to put their chips in production and ultimately bring their products to market, and provides better access to technology. The Yangtze River Delta District, including Shanghai, Zhejiang, and Jiangsu provinces, has the potential to become the center of China's IC industry. A few cities in the area form a critical mass that is likely to fuel continued growth of the industry.

Box 4.2 The Key Factor for Developing Integrated Circuit Design Capacity

When investors started to show interest in China's IC industry, the general view was that design capacity was the bottleneck limiting the industry's growth. However, in 2000 a group of industry representatives pointed to the need for foundries to attract design talent and develop a more complete value chain, including testing and packaging. It was the lack of foundries, rather than design capacity, that was impeding the industry's development.

Following this foundry-first strategy, in April 2000 the Semiconductor Manufacturing International Corporation (SMIC) was founded by Richard Chang, former CEO of a Taiwanese chipmaker. SMIC received heavy government backing and some $1.5 billion in initial investment. Led by SMIC, a large number of foundries have been established in China, including Grace Semiconductor in Shanghai, Huahong in Shanghai, UMC in Suzhou, and TSMC in Ningbo. The opening of SMIC's fifth factory in 2004 was a significant breakthrough for the Chinese semiconductor industry, and China has become one of the world's fastest-growing IC markets.

Because of the industry's momentum, the number of IC design companies in China has also started to grow. In 2000, there were fewer than 100, and most were state-owned companies or device makers with limited design capacity. By 2005 there were over 450 design houses.

Source: IFC 2005.

Despite the industry's rapid growth, a large gap remains between demand and supply, leading to high imports. China accounted for 13 percent of global demand for semiconductors in 2003, up from 7 percent in 2000 (Chase, Pollpeter, and Mulvenon 2004), as more manufacturing in IC-intensive industries—including telecommunications, consumer electronics, and digital television—moves there. With domestic IC consumption growing at almost twice global rates, demand is growing much faster than production capacity. As a result, 90 percent of domestic demand is met by imports; despite the growing number of foundries (see box 4.2), imports are expanding rapidly. In 2004 China imported $55 billion of semiconductors.

The IC design market remains small and at an early stage of development. About 90 percent of the design firms formed in recent years have fewer than 150 employees and 50 design engineers. Many companies with the strongest R&D capability are still developing their first products and have yet to generate revenue. According to CEInet (2004), 90 percent of the chips manufactured in China are exported. A large share of these exported chips end up being re-imported after processing. This means that foreign companies add value to chips produced domestically for consumption ultimately in China, highlighting China's position at the low end of the IC industry value chain.

Furthermore, many top Chinese IC companies focus on low-end applications, particularly smart cards. IC cards have been the low-hanging fruit for Chinese IC design firms for several reasons. First, less design capacity is required to produce a basic IC card, so companies without much in-house R&D capacity can still compete. Second, the Chinese government and major state-owned companies are significant purchasers of IC cards, and in some cases, foreign companies are excluded from bidding on the design work for sensitive government projects. As a result, domestic companies are able to win this less-competitive business. However, margins for basic IC cards are limited, so even a substantial IC card business may not generate enough revenues to fund the development of more advanced products. As a result, some of China's early IC design leaders may find it difficult to climb up the value chain.

Personal Computer Industry

China's computer industry has grown at an extraordinary pace over the past decade. Hardware production jumped from $6 billion in 1995 to $84 billion in 2004, positioning China as the world's largest computer manufacturer in 2004 (figure 4.2).

Most hardware is produced by multinational computer makers—especially those from Taiwan (China). Taiwan's computer makers began moving component production offshore in the early 1990s in response to rising land and labor costs. Taiwanese companies have established large electronics industry clusters in Guangdong province, the Yangtze River Delta, and the Shanghai and Suzhou areas. After restrictions on doing final assembly in China were lifted in 2001, many original design manufacturers from Taiwan moved to the mainland en masse. Production includes manufacturing for desktop and portable PCs, monitors, motherboards, keyboards, cables, and connectors.

Developing and Innovating the ICT Industry

Figure 4.2 Top Global Producers of Computer Hardware, 1995 and 2000–04

Source: Reed Electronics Research 2004.

Local firms dominate China's domestic PC market—especially Lenovo, which in 2004 held a quarter of market share, more than twice that of its nearest competitor (figure 4.3)—and the respective shares of foreign firms such as IBM, Dell, and Hewlett-Packard are decreasing. China's top three PC makers all have their roots in academia. Lenovo, founded in 1984 with funding by the Chinese Academy of Sciences, purchased IBM's personal computing division in late 2004 (box 4.3) and is now the top PC vendor in Asia and the Pacific (excluding Japan). China's second largest computer vendor, Founder Electronics, was formed in 1986 through an investment from Peking University. Tsinghua Tongfang, the third largest domestic

Figure 4.3 Market Shares of the Top Six Personal Computer Firms in China, 2004 (percent)

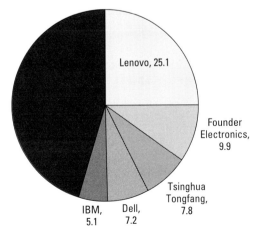

Source: IDG 2005.

> **Box 4.3   Lenovo's Purchase of IBM**
>
> Domestic computer manufacturers have increasingly global ambitions for their business. Lenovo Group Limited, China's top computer maker, bought IBM's personal computing business in December 2004 for $1.75 billion. This makes Lenovo the third largest PC maker in the world (after Dell and Hewlett-Packard) with shipments of 15 million PCs in 2005. The purchase gives Lenovo the opportunity it has always craved to expand beyond China.
>
> Lenovo, founded by Liu Chuanzhi and 10 colleagues with funding by the Chinese Academy of Sciences, began as a distributor of IT products, launched its first PC in 1990, and has led the domestic PC market in China for seven years. Lenovo's success, like that of other large Chinese companies, has been based on the country's vast labor resources and low production costs. However, Lenovo has traditionally carried out little independent R&D and mostly manufactured low-end systems. While this model worked well for the domestic market, where Lenovo has strong ties to the government, its integration of IBM's PC unit will be a challenge in its quest to globalize. As a result of the IBM acquisition, Lenovo now generates more revenue from overseas than China.
>
> *Sources: The Times* 2004; Xinhua Financial Network News 2004.

computer firm and the youngest, launched an initial public offering on the Shanghai Stock Exchange in 1997 with Tsinghua University as the main shareholder.

The PC industry has followed the pattern of other technology markets in China. Foreign firms dominated the market until domestic companies became competitive, at which point foreign firms abandoned the low-end market and tried to shift into higher-value products. Although foreign firms have fared well in the server market, China Langchao is the market leader, and Lenovo is targeting the market.

A major challenge for China's computer industry will be to expand beyond production of PCs for the domestic market and low-value peripherals and components for the global market. Given the country's strong technical skills and ability to attract foreign investment and technology, it has a reasonable chance of meeting this challenge. Lenovo is moving into software, Internet services, and information appliances; other hardware makers are expanding their scope as well.

Software

Software is a key pillar of China's information technology (IT) industry. Both the central and the local governments (such as Beijing and Shanghai) are promoting the industry by facilitating funding for software startups and incubators. For example, Chinese software companies do not have to pay taxes during their first two years of operation and receive a 50 percent tax break in the third and fourth years. The industry also benefits from simplified administrative procedures and relatively fast

Developing and Innovating the ICT Industry

Table 4.1 Chinese Software Parks

Park	Features
Dalian Software Park	• Establishment in 1998 by the Dalian city government • Most tenants are multinational corporations from Japan and the Republic of Korea, as well as local companies
Shanghai Pudong Software Park	• Operations began in 1998 • Joint investment by China Electronic Corporation and Zhangjiang Hi-Tech Park Development Corporation
Beijing Zhongguancun Science Park	• First state-level high-tech development zone, approved by the State Council in 1998 • Largest software development and production center in China, consisting of five science zones
Guangzhou Software Park	• Establishment in 1999 as the software industry base of the National Torch Plan

Source: IFC 2005.

approval of foreign investment. In addition, preferential treatment is given to research facilities that successfully commercialize their research, and local governments have provided financial support for the construction of software parks (table 4.1).

In addition, several aspects of the Chinese economy lend themselves to the software industry's growth, such as the manufacturing sector (which uses software in computer and telecommunications equipment) and widespread use of consumer electronics products and automated machinery, all of which require software bundling. The country's 20 million small and medium-size enterprises provide a substantial business-user base for software products.

China's software industry has grown rapidly in recent years, reaching $25.3 billion in 2004, according to the Software Industry Association's forecast—nearly 30 percent more than in 2003 (figure 4.4). Software products accounted for 51 percent of revenue, system integration for 33 percent, and software and IT services for 16 percent. Software exports, including software outsourcing, are estimated to have totaled $3.2 billion in 2004, accounting for just 1–2 percent of IT industry exports.[3] This reflects China's comparative strength in IT manufacturing and weakness in software.

By the end of 2005 there were 8,000 software companies registered in China (*McKinsey Quarterly* 2005). Most of these companies are private, foreign, and located in Beijing, Guangdong, Zhejiang, and Shanghai. Although many firms have attracted investment from foreign software companies, efforts are generally limited to product localization and customer services. Most domestic software companies are small, and few have developed commercially successful packages. Large players in the market include China National Computer Software and Technology Service Corporation (CS&S), Shenyang NeuSoft, UF Soft, Kingdee, and ZTE Software Co., Ltd.

Features

Three features of China's software industry are notable: its output, which focuses on products; regional dispersion; and lack of pure-play outsourcers (those with more

Figure 4.4 China's Software Market, 1999–2004

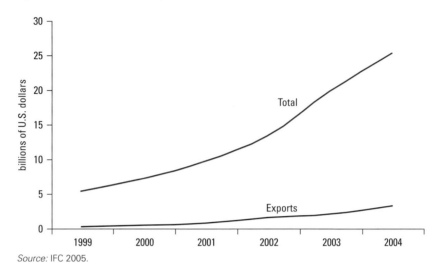

Source: IFC 2005.

than 90 percent of revenues from outsourcing):

- *Focus on products.* Many software firms with strong capabilities and business models have a product orientation, leaving the IT services market underdeveloped. Unlike Europe and the United States—where companies such as IBM Global Services, EDS, Accenture, and Cap Gemini offer a wide range of consulting and systems integration services—China has a limited range of providers of IT services. There are several reasons. First, the government restricts foreign companies from offering many of their services in China. Second, Chinese companies are often reluctant to pay for IT services. Third, China's overall economic reforms have focused on agriculture and manufacturing, rather than services.

- *Regional dispersion.* Regions with extensive high-technology infrastructure have more software firms. Among the country's 25 software districts, the top seven (in terms of sales)—Guangdong, Shanghai, Liaoning, Shaanxi, Jiangsu, Shenzhen (where Huawei is headquartered), and Shandong—account for a large portion of companies, workers, and sales (excluding Beijing, which has many firms). Still, software activity is widely dispersed throughout the nation. Although some cities, such as Chengdu (Sichuan province), do not have many software firms, they have at least one that is large and well-known.

- *Lack of pure-play outsourcers.* The Chinese software-outsourcing industry is highly fragmented and lacks large players dedicated to outsourcing. Instead, the largest Chinese software firms engage in a wide variety of businesses besides software outsourcing—notably big-name companies that have recently diversified from product-oriented software development, such as UFSoft and

Developing and Innovating the ICT Industry

Kingdee. Less-known, medium-size pure-play outsourcers in China have yet to achieve scale. DHC, one of the country's largest pure-play software outsourcing firms, has only 1,700 employees and $33 million in revenues. By contrast, India's top software services firm employs 695,000 people and had revenues of more than $17.3 billion in 2005, with more than 90 percent of resources devoted to outsourcing (Farrell, Kaka, and Stürze 2005). The Chinese software industry is undergoing consolidation with the expectation that the resulting larger firms will be able to better compete with at least second-level Indian outsourcers.

Chinese IT service firms face extensive piracy and intellectual property rights challenges. Although the government has enacted tough anti-piracy laws and conducted publicized crackdowns, piracy rates remain high. About 92 percent of software in China is pirated (IDC/BSA 2003), giving firms little incentive to invest significant R&D resources in the creation of new products. This is a serious concern for China's software vendors and a major barrier to the development of a packaged software industry.

China aims to develop an IT services industry, like India's, that works for foreign firms. In fact, it has attracted a few Indian IT companies to help build its market. However, many Chinese companies lack English language skills, undermining the value of the country's large engineering workforce. In addition, Chinese firms have little or no experience in U.S. or other foreign markets and so have not worked with clients on the front end, writing business process requirements and doing systems architecture and design. The country must develop the long-term human capital required to expand these industries.

Thus China will remain a low-cost location for coding and maintenance and is unlikely to create a software industry that can rival Indian giants for some time. Recently, foreign companies moving their software development operations to China have become a key driver in China's outsourcing market. Whether such development centers are performing software outsourcing in the traditional sense is debatable, since their main customers may be their corporate parents in their home markets rather than third-party customers. Still, such centers create new software export markets and make China a major link in the global software value chain. Expertise developed at these centers will ultimately diffuse into the Chinese market, improving the technical capabilities of domestic software outsourcing firms, which foreign companies may make greater use of in the future.

Information and Network Security Services

Expanding Internet access, as well as escalating network and computer security attacks, are driving demand for information and network security services in China. According to the first survey of information network security, organized by the Ministry of Public Security and the China Computer Association and conducted by

China's Information Revolution

Figure 4.5 Network Security Revenue in China's Vertical Markets, Q1 2003–Q2 2004

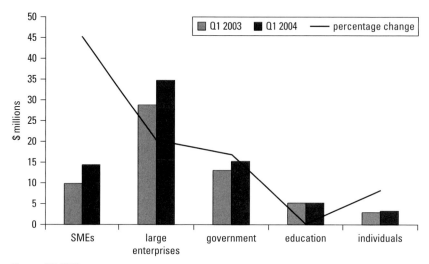

Source: IFC 2005.
a. SME = small and medium-size enterprise.

CRC-Pinnacle Consulting in 2003, 73 percent of organizations reported having information network security issues. About 79 percent of incidents involved computer viruses or worms. In addition, 36 percent of users reported problems with spam, and 43 percent reported unauthorized changes to Web pages. In 2003, the information security market reached $200 million, and by 2007 it is expected to grow to $670 million. The heavily regulated market for government and the more open corporate market are the two biggest markets for information and network security services (figure 4.5).

Although the government does not invest as much in security as large enterprises do—accounting for about 20 percent of the market—government-backed projects have been a key driver of growth for domestic security firms (box 4.4). For national security reasons, China's government relies exclusively on domestic providers of information and network security products. Moreover, a 2003 government procurement law requires investment plans to receive central government approval. The government has been able to use the approval process to ensure that sensitive security projects are handled by domestic vendors.

Large enterprises lead in IT system implementation in China. A 2003 survey by the State Asset Management Commission and China Computerworld Media Group found that 3,000 large enterprises accounted for 21 percent of national IT investments. As IT systems become more complex and critical to businesses, security has become a key issue. Large enterprises have invested in security measures more actively because of their large overall investments in IT. The telecommunications and finance industries lead investments in network security. Since these industries are mostly state-owned, the government encourages them to use domestic security products.

Developing and Innovating the ICT Industry

> **Box 4.4 Leading Domestic Security Firms**
>
> Aided by favorable government policies, Chinese network security firms have made great strides in catching up with foreign firms in terms of technology, services, and marketing. Measured by revenue, five of the country's top 10 network security product vendors were domestic firms in 2004.
>
> TopSec, the leading Chinese network security firm, has received support from the national and Beijing governments and expanded its business rapidly. In 1999 TopSec's Net Guard firewall system became the first such product recommended by the State Security Bureau. Later that year, the Net Guard line of information security software was also included in a list promoted by the government. In addition, Chinese firms have had success with antivirus software, particularly in the consumer market. Rising Technology, Jiangming, and Kingsoft's consumer antivirus products are widely used domestically. Venus Info Tech's flagship product, Cybervision Intrusion Detection and Management System, is a top brand in China's network security market.
>
> Domestic antivirus companies are now developing more advanced products and offering online services. Rising Technology hopes to capture half of the domestic consumer market in a few years, and Kingsoft is aiming for the enterprise antivirus market, investing $1.2 million to develop a proprietary product. Over the long term, domestic companies are more likely to succeed if they have strong technology capacity, continue to invest in security-related R&D, and expand beyond government clients.
>
> *Source:* IFC 2005.

Small and medium-size enterprises (SMEs) are behind in use of network security systems. While foreign enterprises spend 4 percent to 5 percent of their IT budgets on security, most Chinese companies spend less than 2 percent. As the Chinese private sector increasingly adopts internal IT management systems, uses e-mail, and establishes Web sites (see chapter 7), it will likely recognize the importance of network security software (such as antivirus and firewall protection) and reach global norms for investment in security products. The Chinese government is urging SMEs to deploy network protection products amid concern that poorly protected IT systems can be a national security issue. The government has begun inspections to confirm whether companies have implemented security software.

The government's growing concern about information and network security is also reflected in other initiatives. The National Computer Network Emergency Response Technical Team and Coordination Center was established in 2000, and an Internet emergency system was established in 2003. An emergency response system for public computer networks has been developed, and a network security monitoring platform focused on international Internet ports has been constructed. Attempts have also been made to manage e-mail security and control spam.

The Chinese government has also implemented controls on the research, manufacturing, and marketing of security-related products through several certificate approval processes. All such products sold in China must first be certified by the Computer Information Network Security and Product Quality Supervision Center of the Ministry of Public Security. A certificate for information security products from the Information Technology Security Certification Center is also necessary for products sold in the commercial market. For products of intelligence information systems, manufacturers must receive evaluation and certification from the state secrecy authorities. Encryption products must meet required government regulations.

Digital Media

The development of digital media has resulted from the growing integration of IT and traditional content industries. Digital media can be wireless or online—including text messages, games, animation, advertising, learning materials, and other applications—as well as Internet Protocol–based television and radio. Digital media has become a large service market in a short period. Between 2001 and 2005, China's digital media market jumped from $0.5 billion to more than $12 billion (figure 4.6). The industry is expected to maintain high growth over the next 5–10 years and then gradually enter a period of maturity.

Some segments of the digital media industry have grown especially fast. Mobile messaging services—that is, Short Message Service (SMS) and Multimedia Messaging Service (MMS)—attracted more than 240 million mobile telephone users in 2005, up 49 percent from 2004. Markets for Internet games, online advertising, fee-based e-mail services, and search engines have also grown rapidly. The government has

Figure 4.6 Size and Growth of China's Digital Media Industry, 2001–05

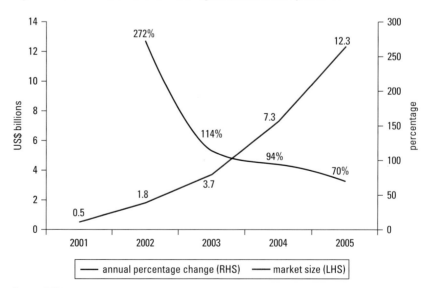

Source: IFC 2005.
Note: LHS = left-hand side; RHS = right-hand side

Developing and Innovating the ICT Industry

formulated a series of policy measures to encourage and support the development of the digital media industry.

Mobile Data Services

The market for mobile data services opened when China Mobile launched Monternet, its wireless value-added service platform, in 2001. Under this platform, service providers offer content and services (such as ring tones, pictures, and news) to mobile subscribers, and China Mobile collects fees on behalf of the providers through its billing system.[4] Service providers pay China Mobile 15 percent of the revenues collected. China Mobile had agreements with over 1,000 service providers at the end of 2004. China Unicom has launched a similar platform, Uni-Info, with a sharing rate ranging from 10/90 to 40/60 depending on the service provider's scale and performance.

SMS is the most popular wireless data service in China. According to the two mobile operators, SMS traffic totaled 1 billion messages in 2000; this number jumped to 220 billion in 2004. All handsets are now SMS enabled, and more than 70 percent of mobile subscribers use it.

Steady growth in the number of mobile data users and applications has made the mobile telephone an extremely popular form of communication and a new type of media in China. As with other media, the government is enforcing strict management. To ensure that only legal mobile data services are offered, the Ministry of Information Industry (MII) (which regulates the wireless sector) issued in 2004 rules regulating the SMS market. The new regulations require operators and service providers to offer a convenient system for unsubscribing from services, preventing users from being charged for unwanted content. The MII's tightening controls over wireless content will have a mixed impact on the market. Transparent regulations will help wash out unwanted or illegal content from the wireless market, providing a fairer playing field for service providers and promoting greater user confidence. However, stricter controls have also led to lower valuations and reduced investor confidence.

While thus far most mobile content has been focused on text-based platforms, 2.5 and 3G networks offer support for more sophisticated high-speed content. The development of SMS-based content and delays in the introduction of 3G mobile networks have limited China's creation of advanced mobile content. However, given the looming introduction of 3G mobile in the country and the large number of mobile subscribers, advanced mobile content could become a significant industry.

Digital Broadcasting

A number of developments are transforming conventional television broadcasting. Television broadcasting is moving from analog to digital, and new methods are emerging to deliver broadcasting over the Internet and to mobile telephones.

China's digital TV industry is at an early stage, with less than 1 percent of the country's 120 million cable subscribers accessing digital programs. Moreover, a number of pioneering city-level cable operators—including those in Shanghai,

Chengdu, Guangzhou, and Suzhou—report that growth in digital television subscribers has slowed. As a result, operators are now subsidizing set-top boxes to attract new subscribers. Cable networks are also extremely fragmented, which is likely to delay growth. The State Administration of Radio, Film, and Television (SARFT) has tried to integrate city networks within provinces, but progress has been slow.

Telecommunications operators are looking to leverage their large investment in broadband networks by offering broadcast services using Internet Protocol television (IPTV). China Telecom and China Netcom have launched trial services with approximately half a million users at the end of 2005. IPTV faces several bottlenecks including regulatory uncertainty, limited bandwidth, and relatively high costs. SARFT has stated that only broadcast companies can provide IPTV, with the result that the telecommunication companies must collaborate with media companies to provide service. Bandwidth on the operator's broadband asymmetric digital subscriber line (ADSL) networks will need to be increased to provide nationwide IPTV service. The service is also currently significantly more expensive than digital cable television.

Despite a variety of competing standards,[5] a number of mobile operators around the world have launched mobile television services. With the large number of television and mobile users, these services could develop into an attractive market for China. One factor that may help drive the mobile television market is the 2008 Olympics, which will be staged in China.

Online Games

China's online gaming market has grown strongly in recent years. Its revenues exceeded the country's movie ticket sales in 2003. In 2004 the market size reached $300 million, according to statistics from the General Administration of Press and Publication. It is expected to be a $1 billion industry by 2008. An online game connects multiple players (sometimes thousands) at the same time. China's online gaming market is being driven by growing Internet and broadband penetration, the increasing availability of PCs and Internet cafes, and low game fees. The development of prepaid card distribution channels by leading online gaming companies allows payment flexibility.

To reduce their reliance on licensed games—about two-thirds of the titles in commercial operation are licensed from foreign companies, mainly in Japan and the Republic of Korea—and avoid sharing revenue with game developers, many Chinese game operators are investing in in-house R&D capacity to shift the market toward domestically developed content. Although domestic game output is increasing, it remains weak relative to leading foreign markets because the development of core technologies (such as network and graphic game engines) has been slow in China. To bring proprietary games to market more quickly, some domestic companies are seeking to partner with or acquire design houses and development teams.

A talent shortage has also slowed the growth of online gaming companies. Experienced game developers, especially project managers, are rare in China. Moreover, available talent is concentrated in a few top companies—more than half of the

Developing and Innovating the ICT Industry

estimated 2,300 game developers in China work for one of the four biggest gaming companies (TQDigital, Shanda, Kingsoft, and Netease). Most leading companies have R&D teams of fewer than 50 people. Weak R&D and shortages of developers will continue to hinder domestic game development.

The government's policies on online gaming are sometimes contradictory, promoting the industry's development while asserting restrictive regulations. This is partially due to the potential negative impact of digital media. Tough regulations have been implemented to punish companies for publishing "inappropriate" content. For example, in September 2004 the Ministry of Culture banned the operation of six online games because of "unhealthy" content. The environment is also complicated by rivalries between the main regulatory bodies over the online game industry. In October 2004 the Ministry of Culture made a formal announcement reemphasizing its authority over the industry. Earlier regulatory announcements had been made by the General Administration of Press and Publication. This situation may continue for some time.

Strategies for Stimulating Innovation and Growth

The demand from China's large domestic market and foreign export markets offers huge opportunities for its ICT industry. Innovation and R&D capacities are critical for the industry to attract investment, maintain high growth, and become globally competitive. China must address the barriers to more effective innovation by introducing regulatory and policy reforms, encouraging collaboration, and increasing government support. It needs a robust national innovation system that includes the following:

- domestic and foreign enterprises that serve as the backbone of technology development, and that are likely to invest in the development and commercialization of new technologies

- research institutes that serve as engines for innovation

- government agencies that can provide strategic direction and create an environment that fosters innovation, including funding and other support measures

- capital markets and venture capital that can provide investment and support for new technology development.

Removing Regulatory Obstacles to New Technologies

The government has an important role in facilitating an environment that incubates a virtuous cycle of innovations in products, processes, and practices. This includes ensuring that the legal framework protects intellectual property rights and relaxing restrictions that may inhibit innovation.

Eliminating entry barriers and allowing more foreign direct investment and joint ventures in the ICT market are fundamental to stimulating domestic innovations. Foreign investment not only supports the technical performance of firms in the short term; it also diffuses advanced technologies to the local economy in the long run.

Organizing R&D around Strategic Technologies

China's R&D capacity must progress much further if the country is to achieve innovations in ICT. China has adopted, absorbed, and used foreign innovations for many years. As a result, it has rapidly developed expertise in manufacturing and become a major producer of ICT products and applications. However, domestic ICT industries rely significantly on foreign technologies, particularly for fundamental technologies. Without greater research in such core technologies, China will continue to lag in the ability to contribute to the development of global technological standards and in the ability to develop its own key intellectual property. Increased R&D and innovation will give China more control over the evolution of its ICT industry, allowing it to respond more rapidly to the dynamic pace of technological flow.

China must also establish strategic directions around technologies critical to its industrial development and future competitiveness. For example, the Republic of Korea invested about $630 million in TDX (Time Division eXchange) telephone switches, light transmission systems, Code Division Multiple Access (CDMA) mobile technology, and DRAM (Dynamic Random Access Memory) chips. Years later, the market value created was about $140 billion, or 220 times initial R&D investments (MIC 2002).

To make best use of its resources and competencies, China should focus R&D efforts on core technologies—products and services that have been identified as important for informatization, necessitate extensive customization, or are unique to the Chinese market or where it has strategic advantages. These include areas such as integrated circuits, network security software, telecommunications equipment, and mobile data applications.

Improving Academic and Business Collaboration

China needs to further integrate enterprises, universities, and scientific research institutes to connect skills development and R&D with industrial development, so that technologies can be turned into practical and relevant applications. Industry-university collaboration brings benefits for both: enterprises can provide universities and research institutes with additional funding for targeted research, while the latter have a pool of highly skilled human resources from which to draw. TRLabs in Canada, for instance, is a leading example of multipartner collaboration (box 4.5).

The government should increase financial incentives for academic and business collaboration. This could include outright funding support, as well as tax holidays and interest subsidies for loans. The measures could create incentives for the dissemination of technology, while contributing to the funding of universities and research institutes.

Developing and Innovating the ICT Industry

> **Box 4.5 TRLabs—An Industry-Led ICT R&D Consortium**
>
> TRLabs, a not-for-profit organization, is Canada's largest ICT R&D consortium and one of the most industry-invested R&D vehicles. Its partners include 11 large telecommunications equipment manufacturers and telephone carriers, 34 small and medium-size enterprises, 4 government agencies, and 5 universities. The organization employs 260 researchers, professionals seconded from industry, professors, and students. It drives the competitiveness of western Canada's ICT industry with the supply of brain power and innovative technologies, while balancing the expectations of its three legs—industry, government, and universities.
>
> Through collaboration in research, development, testing, and applications of ICT, the consortium collectively shares the risks of exploring ICT's new horizons but reaps a much greater reward of ideas, new technologies, and highly qualified people. Since its establishment in 1986, it has created 7 companies and had 74 patents issued, with 78 pending. About 30 percent of the 992 technologies developed at TRLabs have been commercialized by industry. In addition, 50 percent of over 800 TRLab student graduates accept employment with its members and apply their practical skills to support the industry.
>
> *Source:* TRLabs 2006.

Stimulating Investment in R&D

Both government investment and private enterprise R&D investment remain insufficient. China's overall R&D efforts amounted to about 1.2 percent of GDP, much less than in advanced countries (see figure 4.7). On ICT specifically, over the past five years the Ministry of Science and Technology (MOST) invested $400 million in ICT R&D, and the MII provided an additional $60 million in research funds each year. In 2002 R&D investment among Chinese ICT-related firms was 2.4 percent of revenues. OECD firms involved in IT and communications, by comparison, invested 9 percent to 10 percent on R&D.

These amounts are relatively small given China's size. As a result, the number of patents granted internationally to Chinese residents remains remarkably low. The U.S. Patent and Trademark Office granted 1,700 patents to Chinese residents in 2004 compared with 3,554 to the Republic of Korea and 20,173 to Japan (see figure 4.7). Chinese firms must still license many of the core technologies developed and owned by foreign enterprises. Securing additional private investment, including new mechanisms such as capital markets and risk investment, will help increase China's capacity to innovate. In addition, given the success of industry clusters, local governments should increase investment in promoting sector development.

Clear incentives are needed to promote innovations and deploy new technologies. Appropriate tax incentives can stimulate R&D. China could draw on successful tax schemes used elsewhere. Some national tax schemes offer a complete write-off of R&D expenditures made during the year. Others provide this only for expenditures

Figure 4.7 R&D Spending and ICT Patent Applications in Selected Countries, 2004

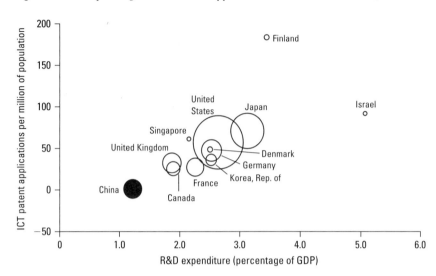

Source: World Bank staff analysis based on World Bank (2006c) and WIPO (2005).
Note: Circles reflect relative size of GDP.

exceeding those of the previous year. Some defer tax payments until enterprises show a profit, an arrangement particularly convenient for startup firms (OECD 1999).

Aligning Standard Development with International Practice

In recent years China has begun to experiment with creating national standards for ICT products. However, relatively few domestic standards or patents have had global success. In 2004 China's ICT-related patents represented just 0.5 percent of the world total.

Despite its large size, China has yet to have a significant impact in international standards bodies. In recent years the International Telecommunication Union (ITU) has reviewed some 4,000 technologies. Of these, just a few Chinese technologies—such as time division-synchronous code division multiple access (TD-SCDMA) and Multi-Service Ring—have been approved by the ITU. Moreover, China has had little input into the Internet Engineering Task Force, and it made no significant contribution to Internet Society image coding standards or to standards set by the Institute of Electrical and Electronic Engineers.

The approval of a standard must also be linked with a commercial strategy to achieve success. In the case of TD-SCDMA, a 3G mobile technology, it has yet to be commercially launched in China or anywhere else. On the other hand, 3G technologies such as wideband-code division multiple access (W-CDMA) and CDMA2000 have been in operation for over half a decade and have a large head start on TD-SCDMA. It is of little use to get acceptance for a standard if the technology resulting from it is not quickly commercialized.

Developing and Innovating the ICT Industry

Some of the Chinese government's proposed standards have attracted international attention but have been opposed by foreign firms and governments. For example, in 2003 the Chinese government claimed that the global Wireless Local Area Network (WLAN) 802.11 standard had security shortcomings and announced that it would establish its own standard, called WAPI (WLAN Authentication and Privacy Infrastructure). To sell in China, foreign companies would have to comply with the WAPI standard and work with licensed Chinese partners, giving them access to detailed product blueprints. Intel (a leading U.S.-based IC manufacturer) balked at this demand and said that it would not manufacture its wireless Centrino chips for notebook PCs under such conditions. The U.S. government became involved, and after eight months, the Chinese government indefinitely postponed (but did not revoke) the WAPI regulations.

Government efforts to develop domestic standards could have a serious impact on Chinese firms—giving them an advantage in the domestic market but making it more difficult for them to be successful in foreign ones if they have to develop different export products based on global standards. As noted earlier, domestic standards could discourage international IT firms from investing in China. A better strategy for China would be to use its large and growing market to influence global standards in ways favorable to its firms. For instance, it could encourage the adoption of open versus proprietary standards to reduce royalty burdens on its domestic companies and allow them to develop technological expertise within open technology platforms. It could also promote the use of innovative IT applications that fit domestic needs, encouraging technology vendors to use China as a test market for new standards.

The ICT market is very dynamic, moves quickly, and is truly internationally integrated. China should be cautious about too strong an involvement in driving the selection of technologies that would be inconsistent with globally adopted standards. This will require patience and maturity to continue to work in and through the international standards bodies to get their technologies recognized and resist the temptation to go independent routes. When it comes to standards, strategic engagement is likely to be more successful than isolationism.

Improving Links between Production and Demand

China faces the paradoxical situation of being a leading global hi-technology equipment exporter but not meeting its own domestic ICT needs. Despite its large IC industry, China still imports a substantial amount to meet domestic demand. Lenovo has emerged as the third largest PC producer in the world, yet only around a fifth of Chinese households have computers. Companies such as Huawei and ZTE have emerged as leading global telecommunications equipment exporters and have sold the latest mobile technology to over a dozen developing and developed nations. Yet China has lagged in introducing 3G mobile networks.

The Chinese government should work with local ICT firms to leverage its significant equipment industry to fulfill domestic needs. For example, it could provide incentives to local computer manufacturers to develop low-cost PCs to increase penetration among lower income households. This could have tremendous export potential since many developing countries face the same challenge of being unable to afford ICT products. The large Chinese market means that producers can achieve significant economies of scale with consequent lower prices. The government should also move more quickly to remove obstacles to the rapid launch of new technological products and services into the Chinese market. This would allow leading-edge technologies to be implemented soon after standards approval, giving China a head start over other countries. Implementation of these measures would produce substantial benefits, enhance sales, and establish China as a showcase for new technology.

Notes

1. People's Daily Online. http://english.people.com.cn/200503/23/print20050323_177881.html.
2. See http://www.oecd.org/document.
3. Japan accounts for 61 percent of Chinese software exports, followed by the United States.
4. This has helped spawn a mobile data software industry lead by companies such as Sina, Tom Online, Sohu, Netease, and Tencent.
5. Competing technologies include Digital Multimedia Broadcast, Digital Video Broadcast-Handheld, and MediaFLO, a proprietary technology developed by QUALCOMM.

Chapter 5

Improving ICT Human Resources

Human capital is the driving force of an information society. Three levels of information and communication technology (ICT) human resources are critical to informatization:

- a general public able to use ICT applications at work and home
- informatization managers who lead ICT development in government and business
- ICT professionals experienced in network design, software development, and research and development.

China's compulsory basic education has dramatically increased literacy—to more than 90 percent in 2003. Gross enrollment in secondary education rose from 46 percent in 1980 to 70 percent in 2003, while tertiary enrollment increased from 2 percent to 15 percent (figure 5.1). Nonetheless, China's enrollment ratios remain much lower than those in developed countries and other BRIC countries (Brazil, Russian Federation, India, and China), except India.

China has a big shortage of the skilled ICT workers needed to implement its informatization strategy and maximize the strategy's economic impact. It is essential to quickly develop human resources of professional caliber. It is also necessary to raise awareness about the importance of informatization and encourage wider participation by the public.

ICT Education and Training

ICT capacity can be developed both at China's formal education system (primary and secondary schools, universities, and technical institutes) and through training and programs outside the regular school system (vocational and certification programs and distance education).

Figure 5.1 Gross Secondary and Tertiary Enrollment Ratios in Selected Countries, 1980 and 2003 (percent)

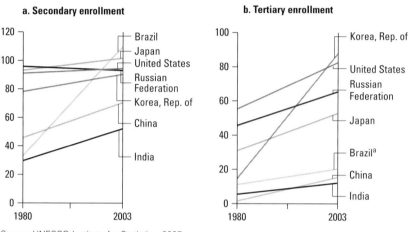

Source: UNESCO Institute for Statistics, 2005.
a. Data from 2002 were used for Brazil.

Primary and Secondary Education

Informatization education should start in primary and secondary schools to cultivate the basic abilities—such as using technology and searching for information—essential to an information society. As early as the 1980s, developed countries such as France, Japan, and the United States included information technology (IT) in the curriculums of primary and secondary schools. Students are taught ICT and to use it as a tool to aid learning in other subjects. Developed countries have substantially increased investment in ICT infrastructure and content in basic education, keeping pace with fast-changing technologies.

By 2010 the Chinese government aims to provide 90 percent of elementary and secondary schools with Internet access and for them to offer at least one ICT course for all students. Since the late 1990s, the government has spent $6.0 billion to $7.5 billion a year on information infrastructure in these schools. According to the Ministry of Education, by April 2004 intranets had been created in 35,000 primary and secondary schools—10 times more than in 2000. However, access to ICT still lags in such schools. In 2004 computer coverage at primary and secondary schools was 36 percent, and only 16 percent of teachers had taken ICT training. There are large regional gaps. In 2001 the ratio of students to computers was 17:1 in Shanghai and 186:1 in Yunnan (Li 2005). Interestingly, regional differences exist in personal computer (PC) penetration rates at primary and secondary schools but not in ICT training for teachers (figure 5.2).

As a result, many students in China lack basic ICT skills (box 5.1), which could have a profound impact on their ability to analyze information and solve problems.

Improving ICT Human Resources

Figure 5.2 Regional Differences in ICT Education in China's Primary and Secondary Schools, Selected Provinces

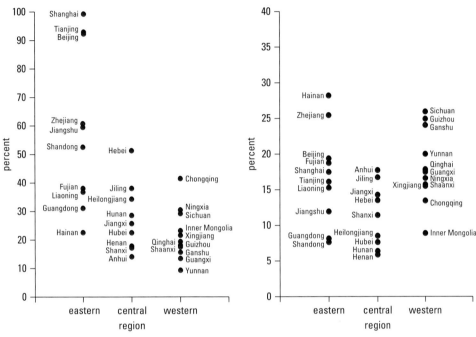

Source: World Bank 2006a.

Universities and Technical Institutes

Strengthening ICT education at universities and technical institutes is an important step to produce ICT professionals, train ICT faculty, and develop highly educated workers with ICT skills. In recent years many Chinese universities have adjusted their programs and curriculums, resulting in new courses and increased enrollment

---

Box 5.1 Information Retrieval Abilities among Primary and Secondary School Students

A recent survey in Shanghai found that primary and secondary school students' ability to obtain useful information from the Internet was low. Only 45 percent of the students used search engines (such as Sina, Sohu, and Yahoo! China) to look for information, while most did not—and 7 percent did not even know what a search engine is.

Most students got Web site addresses from fellow students. Often when they were not satisfied with the search engines they were using, they did not know how to choose better ones (such as Google). Information retrieval did not seem to be part of school curriculums.

Source: Rong and Li 2002.

in ICT-related majors. For example, for the development of IT experts with practical skills, a pilot graduate school of software and microelectronics was established at Beijing University in 2002. Faculty for the school are recruited internationally, including Chinese returnees (who have studied abroad or worked in foreign high-technology companies) and foreign Ph.Ds. The school's curriculum was developed with advice from leading multinational IT companies such as Microsoft, Lucent, and Motorola. The school enrolls about 2,000 students who, in addition to taking classes, conduct research and spend year-long internships in industry (Penrose 2005).

Links between classroom teaching and real-world applications remain weak at most Chinese universities and technical institutes. Lectures by industry experts and researchers could be used to teach ICT majors about new technologies, changes in the information environment, information exchange mechanisms, intellectual property rights protection in a networked environment, and developments in ICT.

ICT education in universities and technical institutes should not only teach basic theories and research methods but also emphasize their application. Practical training and internships at ICT units in companies or government agencies can give students the opportunity to get hands-on experience, learn skills needed to solve real-world problems, and align their learning with market demand. To that end, many Chinese universities and technical institutes have started to cultivate stronger links with enterprises. Such collaboration can provide mutual advantages: enterprises have funding and advanced equipment, while universities and scientific institutes have pools of potential experts and future talent.

Vocational Training and Certification Programs

In China, training outside the regular school system helps make up for shortcomings in the education system, enabling ICT workers to update their skills and narrowing the gap between demand and supply. This type of training is typically provided through centers that are often linked to experienced IT firms.

Because the training system has developed quickly within a short period, it has some weaknesses—mainly involving uncertainties about the quality. The number of training certificates issued in recent years has become excessive, and many students take ICT training to acquire certificates so that they can find good jobs rather than to improve their skills.

To gauge ICT skills, the Chinese government conducts several major exams, such as the National Computer Rank Examination (NCRE). In 1994 the Ministry of Education introduced the NCRE to test and rank knowledge of computer applications, and by 2001, more than 2.15 million people had taken it (You and Tao 2003). However, the exam has lagged behind market needs, and students who have passed it are not necessarily capable of using modern ICT.

More recently, internationally recognized technical certifications have also been used in China, including for operating systems (Microsoft, Linux, and Unix),

Improving ICT Human Resources

> **Box 5.2 IBM's Software Engineer Training and Certification Program**
>
> To develop skilled workers for Shanghai's software industry—and support its goal of becoming China's biggest software export base—the municipal government signed an agreement with IBM in 2003 to train software engineers. IBM and the Shanghai Shenxin Information Technology Academy (which is affiliated with the Shanghai Informatization Office) provide a range of ICT training and certification to 250 teachers and 5,000 software programmers and engineers in Shanghai, Jiangsu, and Zhejiang over a three- to four-year period to meet the demands of the growing software industry. In addition, honor students from the program will be offered internships at IBM and its partner organizations.
>
> *Source:* http://tech.tom.com/Archive/1121/1793/2003/7/30-54181.html.

networks (Cisco, CIW, and 3Com), servers (IBM, Sun, and HP), database management (Oracle and Sybase), and graphic design (Adobe and Macromedia). China is further strengthening its ICT training in collaboration with multinational IT firms (box 5.2) and academia.

Distance Education

To supplement these efforts and further cultivate ICT skills, China is expanding distance education, which has been an important component of its growing education market. In 1999 the Ministry of Education approved 67 universities, including the Central Broadcasting and Television University, to provide distance learning. By 2003 more than 2,100 distance learning centers had been established outside universities, serving more than 2.2 million students.

Obstacles to ICT-supported education are more severe in China's rural areas—particularly in western and central regions, which also suffer from low literacy rates. In response, in 2003 the State Council increased financing for distance learning in rural primary and secondary schools, especially those in western and central regions (X. Li 2004). Distance learning can help narrow the digital divide and ICT education gap between these regions and those in the east, as well as between rural and urban areas overall. It can also help reform rural and traditional (nondistance) education institutions.

Distance learning programs should equip classrooms with IT equipment, long-distance communication facilities (such as satellite connectivity), and multimedia learning materials such as CD-ROMs and videos. In addition, knowledge sharing among education institutes should be encouraged to improve the effectiveness of distance learning (box 5.3).

> **Box 5.3** **Launch of the Union of National Teachers Education Network**
>
> Distance learning enhances the skills of teachers in rural areas, benefiting students. In 2003 the Ministry of Education launched the Union of National Teachers Education Network to improve teacher skills and established a lifelong learning platform. The network focuses on improving the quality of rural teachers and sharing knowledge among them by providing education resources and common application platforms. In addition, the network helps identify the most effective methods for training rural teachers and courses for distance education.
>
> *Source:* Chinese National Commission for UNESCO 2004.

Challenges

China has enormous demand for ICT skills and faces several challenges in providing the education and training to develop these skills.

Lack of Public Awareness and Application Capacity

Limited awareness of informatization undermines the public's interest and ability to use ICT. China's success in informatization will be determined by its people's willingness and ability to acquire information, share knowledge, and use ICT effectively. Public awareness is also critical to build support for China's informatization policies.

The ICT capacity of the general public is important because it affects the absorption rate of ICT applications. According to the Chinese Network Information Center's semiannual Internet survey, the main reason Chinese people do not use the Internet is because they do not understand computer and Internet technologies (figure 5.3). Another factor contributing to low public awareness is that access to ICT—especially broadband Internet access—is still limited in many parts of the country (see chapter 3).

Broad public support is a key element of the successful implementation of China's informatization strategy; the absence of such support could foster dissension over the most efficacious allocation of investment resources.

Lack of ICT Management Expertise

As e-government and e-business continue to develop (see chapters 6 and 7, respectively), ICT management expertise has become ever more crucial. Yet many Chinese firms and government agencies lack the managerial understanding and skills required to integrate ICT applications. Managers may not know their options for ICT solutions and often do not have the information they need to make decisions about the allocation of resources and effort regarding ICT equipment and human resources.

Improving ICT Human Resources

Figure 5.3 Chinese People's Reasons for Not Using the Internet, 2005

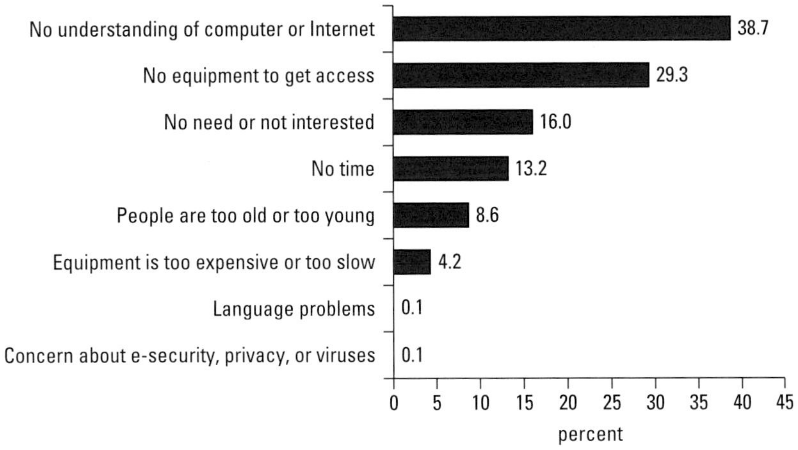

Reason	Percent
No understanding of computer or Internet	38.7
No equipment to get access	29.3
No need or not interested	16.0
No time	13.2
People are too old or too young	8.6
Equipment is too expensive or too slow	4.2
Language problems	0.1
Concern about e-security, privacy, or viruses	0.1

Source: CNNIC 2006.
Note: Findings are from a survey question that allowed multiple responses.

In many organizations the internal IT division or information center is the main source of ICT skills and knowledge. Managers of these divisions provide technology solutions and supervision but are often not involved in strategic decision making. Mid-level managers do not have authority over other divisions, making informatization planning, coordination, resource allocation, and promotion extremely difficult. In addition, most of these IT managers are engineers with technical knowledge but without modern public or business administration experience. As a result, some e-government and e-business projects in China over the past decade did not have clear impacts despite huge investments in information systems and business reengineering applications.

Recent initiatives, such as that of the China Information Association, are targeting the creation of chief information officer (CIO) structures in both the public and the private sectors. Related on-the-job training in such areas as e-government, informatization planning, and project management is being provided to government leaders, members of the State Council Informatization Leading Group, and IT managers. The main goal of the training is to build recognition among decision makers that having a CIO to manage informatization—that is, ICT use in production and sales processes—integrates flows of information and physical materials, optimizing resource use and processes and enhancing efficiency and competitiveness (box 5.4).

Another goal of these initiatives is to develop a pipeline of potential CIOs and other ICT managers among university graduates, combining technical skills with business and public administration skills. For example, Tsinghua University's graduate-level enterprise informatization course, introduced in November 2003, is designed to develop management expertise in enterprise informatization and has

> **Box 5.4 Training for Government CIOs**
>
> The first training session for government CIOs in Ningbo took place from July 30 to August 1, 2004, in Xiangshan. The seminar was attended by 82 leaders and information officers from 56 organizations. Officials from the State Council Informatization Office, renowned foreign and domestic ICT specialists, and informatization experts from Ningbo were invited to give lectures.
>
> Among the topics covered were trends in informatization development and corresponding policies, the central government's informatization strategy, and the overall plan for developing ICT human resources. Participants discussed e-government in depth. Through the training, they learned more about the increasingly important role that CIOs play in management decisions affecting government informatization.
>
> *Source:* http://www.nbit.gov.cn.

attracted individuals with ICT management experience, including CIOs, chief technology officers, and mid- and high-level managers (*Microcomputer World* 2003).

Supply-and-Demand Gap for ICT Professionals

Due to different concepts, sources, and definitions, statistics about the number of ICT workers in China conflict, making it difficult to analyze trends and forecasts in the market for ICT employment. Nonetheless, available information suggests that a large gap exists between the supply and demand for ICT labor.

According to one definition, about 4 million workers were employed in China's information industry in 2003. This was 11 percent more than in 2002—much faster growth than in national employment (0.9 percent) and urban employment (3.5 percent). Among these workers, 3.5 million were in the manufacturing industry (7 percent more than in 2002) and 0.6 million in the software industry (18 percent more) (MII 2004).

These 4 million workers accounted for less than 1 percent of China's labor force—an extremely small share, especially relative to developed countries. According to the U.S. Department of Commerce, half of U.S. employees will work in IT firms or for information-intensive employers in 2006. The European Union estimates that 80 percent of its workers will be in the services industry after 2005, mostly in IT-related jobs (China Dong Wu Ren Cai Hotline 2005).

China suffers from large gaps between the availability of and demand for ICT professionals with masters or doctorate degrees. For instance, in 2002 there was a need for about 80,000 more such professionals than were recruited in 1999 (table 5.1). The need for ICT professionals will only grow as China moves toward an information society. Although the number needed to fulfill the 11th Five-Year Plan will likely be much higher, Chinese universities and institutes will produce only 40,000 students with an IT or IT-related major in each of the next five years, according to the Ministry

Improving ICT Human Resources

Table 5.1 Annual Supply of and Demand for ICT Professionals in China, by Field

Indicator	Electronics and communications engineering	Computer science and technology	Economic management
Recruited masters or doctorate degree–holders, 1999	4,200	4,300	22,500
Demand for such recruits, 2002	23,600	8,000	81,400
Gap	19,400	3,700	58,900

Sources: Ministry of Education 2003; NBS 2003; NBS and Ministry of Science and Technology 2003.

of Education. The gap is expected to be especially wide for software development. There are only around 200,000 software specialists, and training capacity is limited to some 15,000 a year (Yuan 2006).

Brain Drain

The outflow of human resources to more advanced areas or countries constrains economic development in many developing countries. As of 2002 China had sent 580,000 students to foreign countries. Only 150,000, or 26 percent, have returned (S. Li 2001). This is below the average return rate of 35 percent for developing countries.

This trend is amplified by the migration of skilled labor from China's western to eastern regions and from rural to urban areas. Between 1982 and 2002, the country's urban population increased from 191 million (19 percent of the national total) to 502 million (39 percent). Not only are residents of more developed regions unwilling to work in underdeveloped areas, but few graduates return to rural areas once they get their urban residency (through such means as college education or military service).

Brain drain is not unique to China, nor is it an outcome of informatization, but it does affect informatization in several ways. On the one hand, successful informatization requires a trained and skilled workforce—precisely those who tend to emigrate. On the other hand, informatization can help mitigate brain drain by providing novel training opportunities such as distance education.

One reason for the migration of skilled labor from China's rural areas is that awareness of the significance of informatization for economic development is not yet well established in underdeveloped areas. Low awareness leads to low investment in ICT infrastructure, which impedes further investment in ICT programs. All of these factors drive away educated, technologically savvy residents from underdeveloped areas.

By contrast, more developed provinces and cities—particularly municipalities directly affiliated with the central government—attach great importance to informatization and take active, effective measures to train and retain ICT workers. In addition, wage levels and working environments are much better in developed areas.

Movement Forward

As informatization accelerates, China will need more ICT experts as well as workers with adequate ICT skills. It has become a pressing task to improve and expand the human resource base to meet the demand for the country's ICT development. The various stakeholders—governments, enterprises, schools, research institutes, and individuals—must work together to build up the human resources needed to maximize the economic and social benefits from ICT development. It is also important to create a shared vision among and within stakeholders to coordinate plans and projects for educational informatization.

Informatization requires different levels of human capacity for different purposes and at different stages. The numbers and skills of technical experts should be defined in the context of industry development trends, balancing supply and demand. In that respect, there needs to be a better understanding of the market for ICT skills, ongoing official forecasts of supply and demand, and regular monitoring of the situation to minimize gaps. Structural changes to government administrations and business enterprises are needed to incorporate ICT as a strategic function and to train ICT managers. China's governments, enterprises, and schools must work together to increase ICT knowledge and skills at all stages of people's lives to enable them to participate in informatization.

Improving Educational Access and Quality through ICT

Educating and training China's vast population is a challenge. ICT diversifies the range of learning opportunities. It helps overcome physical shortages of teachers and classrooms through online education delivery. ICT also provides new and innovative ways of teaching that can make learning more interesting while offering students an opportunity to enhance computer skills.

China has offered distance education by television and radio for more than 40 years, helping to meet the challenge of extending schooling beyond the primary level. For tertiary education, the Internet's capacity for two-way interaction offers the promise for improving access and affordability and for providing flexibility to combine work with further study. Nearly one million students are studying online in China, many taking classes from developed country universities without having to migrate (Perkinson 2005). The challenge is to raise the quality and accessibility of distance education while maintaining the low cost.

The Chinese government has made substantial efforts to provide ICT-supported education. The China Education and Research Network (CERNET), the country's backbone infrastructure for education, is the world's largest online education and scientific research network. It is also among the top three academic networks in the world, enabling 70 percent of Chinese universities to provide distance learning through their intranets. CERNET reaches more than 200 cities, including 36 provincial capitals. Some 1,300 organizations, including 800 universities, access the Internet through CERNET, and 15 million users access CERNET through 5 million PC

Improving ICT Human Resources

terminals (Ge 2004). CERNET's capacity could be further enhanced by engaging the private sector.

China has initiated a number of projects to enhance distance education and other ICT infrastructure in rural areas. These have included a $108 million project to connect universities in the western part of the country to CERNET and $100 million for ICT infrastructure in rural primary and secondary schools. Nonetheless, there is still a significant lag between urban and rural schools. Greater investment is needed in rural areas to provide information infrastructure, train teachers, and develop appropriate ICT applications.

China is also creating digital campuses, with school intranets that provide services to students and administrators; multimedia classrooms that offer digital recording facilities, electronic blackboards, and Internet access; and multimedia libraries that provide access to online information resources and interactive courseware. These efforts should be intensified with greater industry collaboration.

For example, countries ranging from South Africa to the United States have universal service policies requiring telecommunications operators to provide Internet access to schools. Another model is encouraging computer manufacturers and software companies to provide free or discounted products for educational users. China has had some success with the latter approach, engaging both multinational and domestic firms such as Apple and Lenovo to provide products for education. Too often, however, these projects have been one-off. These types of programs should be institutionalized with ongoing monitoring of targets for school connectivity.

Raising Public Awareness

A widespread program is needed to build public awareness of ICT, including ways to access and use it and its role in economic and social development. Efforts are needed to promote e-government—to encourage citizen use of online public services and participation in decision making. One way of achieving this is by putting popular public administration applications online (such as for obtaining licenses or paying taxes). Public awareness of ICT can also be increased through media campaigns on the radio, television, and Internet (box 5.5).

Programs also need to be established to provide widespread ICT training to the general population. This is particularly critical for people who are unlikely to be exposed to ICT through the educational system or work, including groups such as homemakers and persons who are unemployed, disabled, or elderly. One example of a successful policy in this area is in the Republic of Korea. In 2000 the Korean government established a computer educational plan targeted at marginal users. By the end of 2003, the program had trained over 20 million people, including persons who are disabled, elderly, or have low incomes, as well as women and rural residents.

Public facilities with access to ICT are needed for people without access at home, school, or work. E-government can be promoted by increasing the number of Internet kiosks in public buildings, enabling individuals to access local, national, and international information resources. The government should consider a widescale

> **Box 5.5 Beijing Raises Public Awareness of Informatization**
>
> As e-government began expanding in Beijing in 2002, the city conducted a campaign using brochures, the media, face-to-face training, and online courses to raise public awareness and recognition of its fast and convenient online public services. In addition, professional training institutes, contracted by the city government, provided applied ICT training to the public for free. More than 30 classes were given in different locations throughout the city, and over 4,000 firms and individual users were trained in the basics of using online government services. In addition, by January 2003, 48,000 civil servants in Beijing had passed the e-government certificate examination.
>
> *Source:* State Information Center and China Information Association 2003.

program that would ensure access to e-government applications—including training and assistance—through a network of public facilities. For example, public access facilities are an essential component of Mexico's program to ensure that all citizens can benefit from e-government. The Mexican government has analyzed areas where there is a lack of access to ICT in order to develop digital community centers. More than 7,000 centers have been established in schools, libraries, health centers, post offices, and government buildings.

Mainstreaming ICT in Education and Training Systems

Requiring ICT classes in primary and secondary schools is essential for building basic skills among the population. Often there is too much emphasis on connecting schools to the Internet without enough attention paid to teaching ICT skills. Schools need to provide teachers with ICT training—including general competency in ICT use and, more important, appreciation of when and how to mix ICT and other media into standard pedagogies. The Chinese government should create incentives to mobilize private participation in education investment and improve school infrastructure for ICT-supported learning (for example, by providing and maintaining equipment).

Over the medium and long term, China needs to add ICT-related majors in higher education—or even launch local ICT institutes—to bridge the gap between supply and demand for ICT-skilled labor. India, for instance, has adapted to the requirement for labor for its IT-enabled service sector by establishing Institutes of Information Technology. The first was established in 1998 through a government-industry partnership. Most have followed that public-private partnership model, with some becoming flagships for multinational firms such as Microsoft, Oracle, and IBM, as well as large domestic IT firms.

The guiding principles of modern ICT education include multiple levels tailored to different audiences, responsiveness to industry trends and market demand, flexible teaching materials easily customized to student needs, and an emphasis on the applicability of skills. Such courses can teach lower-level university students how to improve their ability to retrieve and apply information. For upper-level students, ICT skill development should be integrated with their areas of specialization, promoting the ability to analyze information related to their majors.

In addition to mainstreaming ICT in the formal education system, it is important to normalize and expand training and certification programs to enable people of all ages to enjoy ICT education according to their own demand. Enterprises can equip the labor force with ICT skills through on-the-job training, vocational training, and online courses.

Strengthening University-Business Links

University-affiliated enterprises are common in China. Indeed, most big ICT companies were spun off from universities. For instance, Lenovo (the world's third-largest computer vendor) originated from the Chinese Academy of Sciences and the Great Wall Group (a leading manufacturer of computer parts and provider of system integration and broadband services) from Beijing University. Yet these are rare successes, and innovation capacity in China is still low relative to that of developed countries (see figure 4.7). The linkages between research institutes and enterprises are often not very effective, and few innovations are being commercialized.

University-enterprise partnerships provide a win-win relationship for promoting ICT skill development. Universities can get additional funds for innovation and have the opportunity to steer their technology research to meet practical application demands. University-affiliated companies help commercialize new products and services and make profits that can be used to promote additional research. Such collaborations also give students internship opportunities at ICT companies, allowing them to gain industry experience.

The Chinese government should strengthen the links between universities and firms by providing institutional foundations such as the establishment of intermediary institutions, as has been done in developed countries such as Germany. At the same time, the government should enforce intellectual property rights more strictly to support incentives for inventors (see chapter 4).

Establishing Multidisciplinary, High-Level Information Management

Most government organizations and firms have not institutionalized ICT positions at a senior level, undermining their ability to use information as a strategic tool. One way of enhancing the visibility of ICT within an organization is by establishing a CIO as a senior position. Under this approach the CIO function is elevated to a high level, with associated power, prestige, and pay. CIOs need to be able to influence company decisions and understand the company's strategy, customers, markets, and world trends that affect the firm.

CIOs should be made responsible for leading government and enterprise informatization and spreading ICT for internal efficiency and external service delivery. Appropriate management of ICT adoption requires personnel who can combine business and technology disciplines to accomplish the following:

- select from a broad range of ICT applications (from basic e-mail to more sophisticated applications such as data exchange or supply chain management) that save time and resources in business processes and strengthen the core competencies of the government agency or firm

- provide guidelines on which types of ICT technologies and services are most appropriate for their agency, firm, or industry

- avoid incompatible systems and networks among firms, clients, suppliers, and regions

- maintain and upgrade ICT applications to fit business needs and technology changes.

Government agencies in China have begun to adopt the CIO approach. One example is the General Administration of Customs. Such efforts need to be expanded in both government and industry to more fully realize the benefits of ICT.

Chapter 6

Advancing E-Government

ICT applications make the flow of information more efficient and systematic, supporting the development of an information society. Currently, however, China's 111 million Internet users account for less than 9 percent of its population. Several obstacles impede the spread of Internet use and other information and communication technology (ICT) applications. First, many municipalities and local governments are unable to provide the public with affordable, convenient information access points. Second, significant ICT demand remains latent due to low awareness levels and usage rates. Third, domestic ICT providers are unwilling to cater to users given the seemingly modest demand. Until demand and supply are synchronized, ICT growth will remain slow.

Government can play a significant role in stimulating ICT demand and supply. The Chinese government is probably the country's largest investor in ICT and can lead the adoption and use of ICT applications in several ways. One is for the government to become more capable with information technology (IT) and e-enabled. The government can have a tremendous impact on the ICT sector as a user, purchaser, and provider of ICT services. E-government initiatives make public administration more efficient and transparent. It can also provide citizens with direct access to public services, improving interactions between officials and citizens.

E-government in China has developed in three stages (figure 6.1):

- incorporating ICT applications into internal government processes
- using such applications to improve administrative and management capacity
- introducing e-government applications to deliver public services.

Although different administrations and departments are at various points in these stages, the awareness of advancing e-government has increased in recent years. General capacity has moved beyond administrative applications and into public service applications, and the development of e-government has made noticeable

Figure 6.1 Three Stages of E-Government Development in China

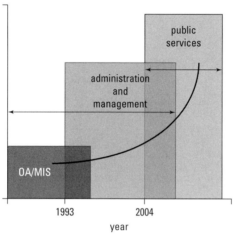

Source: Author.
Note: OA/MIS = Office Automation/Management Information System.

achievements: The Golden Projects are playing an important role in improving governance and the delivery of public services. Increasingly, government Web sites have become a window for government departments to serve the public. Overall, e-government in China is entering the third stage characterized by intensifying applications and services.

Incorporation of ICT Applications into Internal Government Processes

China began incorporating IT applications and networks into government processes in the mid-1980s. It has since expanded the use of local area networks (LANs)—which, among other things, enable e-mail communications and Internet access—into all government departments. It has also built a basic e-government platform, consisting of metropolitan area networks and wide area networks that connect the central government to deputy provincial governments, and LANs that connect 47 governments above the deputy provincial level.

Government departments have generally adopted internal informatization appropriate to their functional objectives (figure 6.2). The most widely used applications involve financial management systems (83 percent), network security systems (76 percent), Web site development (75 percent), and office automation (69 percent).

Major Public Service Projects

The Chinese government has expressed the need for using ICT to increase administrative transparency, enhance management efficiency, and promote honest government. Government informatization initiatives aim to simplify administrative

Advancing E-Government

Figure 6.2 ICT Application Use by Chinese Government Departments, 2004

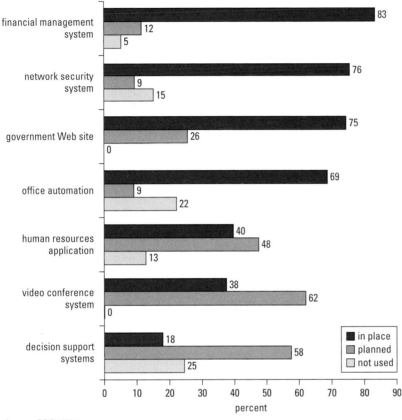

Source: CCID 2005.

procedures, cut transaction costs (such as registration fees), and improve public services for the average citizen. Continued use and expansion of e-government services will largely depend on public satisfaction with them. Thus, the government has an important role to play in setting standards for the services it delivers using ICT applications.

The guidelines for China's e-government development, issued by the State Council Informatization Office in 2002, emphasize two networks (intranet and Internet), one Web site (government portal), four databases (demographic, juridical, geographic and natural resources, and macroeconomic data), and 12 "golden" projects. A 2003 survey found that 98 percent of responding government organizations had deployed some type of e-government project.[1] Still, China ranked 57 of 179 countries on the United Nations e-government index in 2005, up 10 positions from 2004 (see table 6.1). This ranking suggests that while China continues to improve in terms of the number of government Web sites, it lags behind many countries on the number of e-transactions and the level of intra-government connectivity (UN 2005).

Table 6.1 E-Government Readiness Rankings in East and South Asia, 2004 and 2005

Country	Index 2005	Global rank 2005	Global rank 2004	Change
1 Korea, Rep. of	0.8727	5	5	0
2 Singapore	0.8503	7	8	1
3 Japan	0.7801	14	18	4
4 Philippines	0.5721	41	47	6
5 Malaysia	0.5706	43	42	−1
6 Thailand	0.5518	46	50	4
7 China	0.5078	57	67	10
8 Brunei Darussalam	0.4475	73	63	−10
9 Mongolia	0.3962	93	75	−18
10 Indonesia	0.3819	96	85	−11
11 Vietnam	0.3640	105	112	7
12 Cambodia	0.2989	128	129	1
13 Myanmar	0.2959	129	123	−6
14 Timor-Leste	0.2512	144	174	30
15 Lao PDR	0.2421	147	144	−3

Source: UN 2005.

Golden Projects

The Golden Projects, most of which were introduced in the late 1990s, have facilitated China's transition from a strictly administrative use of ICT applications to a more public service–oriented use (table 6.2). The scope of the projects varies, from automating internal processes to creating interactive Web sites allowing online transactions.

Implementation of these projects has been widespread and is largely complete (figure 6.3). For instance, the three primary applications of the Golden Shield project—a household registration management system, a crime case management system, and a 110 police alarm system—are now used by all government departments. The project also introduced a safeguard and antitheft system and a geographic information system. Because of the Golden Wealth project, nearly 85 percent of government agencies now use the electronic budget management system. The project also introduced ICT applications for financial payment management, salary payments, and budget accounting, with adoption rates ranging from 69 percent to 77 percent.

Some of the Golden Projects have produced impressive results. For example, 13,874 fake customs declarations were discovered in 1998, amounting to more than $11.2 billion in undeclared exchange. Since the introduction of an e-customs system in 1999 under the Golden Gate project, smuggling has been reduced by curtailing false customs declarations, and the negative effects of illegal currency exchanges on tax revenue have been mitigated by better foreign exchange management. The

Table 6.2 China's Golden Projects

Project	Stakeholder	Description
Tier one (the most high-profile)		
Golden Gate	Ministry of Foreign Trade, Customs, Jitong Co.	Develop an information network of foreign trade activities to speed up customs clearance and enhance the authorities' ability to detect and prevent illegal activities, and collect taxes and duties
Golden Bridge	Ministry of Electronics, State Information Center	Create an infrastructure backbone for the first national economic information network
Golden Card	People's Bank of China, Ministry of Electronics, Ministry of Internal Trade, Great Wall Computer Co.	Create a unified payment settlement system to enable the wide use of credit and debit cards
Tier two (designed to apply information networks to speed up economic reform)		
Golden Macro	China Ex-Im Bank, Ministry of Finance, State Information Center	Strengthen the central government's (Central Economic and Financial Leading Group) macroeconomic control over national economic activities and analysis capacity
Golden Tax	Ministry of Finance, Ministry of Electronics, National Tax Bureau, Great Wall Computer Co.	Computerize the tax collection system, prevent tax evasion and fraud, and reduce tax losses; also allow customs departments to verify a range of data to facilitate customs management
Golden Wealth	Ministry of Finance	Construct a government financial management information system to support budgeting, payment and accounting
Tier three (sector-specific IT applications)		
Golden Agriculture	Agriculture	Build a databank service network providing agricultural information, weather reports, and market information
Golden Audit	National Audit Office	Transform the reactive audit system to a more proactive one by having a secure platform for information sharing and for tracking and checking accounts
Golden Quality	Government departments	Enhance regulation, quality, transparency, and service orientation of government units that carry out quality control activities such as certification of products and services
Golden Social Security	Ministry of Labor and Social Security	Better manage the increasing labor force covered by the national insurance system and offer retirement and medical information to the insured
Golden Shield	Ministry of Public Security	Strengthen central police control, responsiveness, and crime-fighting capacity
Golden Water	Ministry of Water Resources	Harness technology for the collection, transfer, storage, and management of water resources; also encompass a system to better manage floods and droughts in the country

Source: Yong 2005.

Figure 6.3 Implementation of Selected Golden Projects, 2004

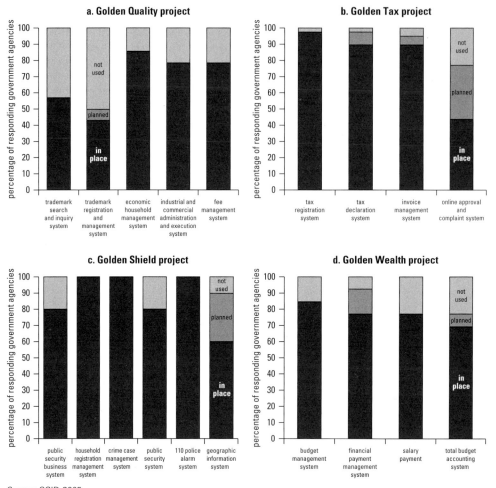

Source: CCID 2005.

e-customs system enables private enterprises to electronically declare customs information, track export permits and vouchers, and provide inspections data. The system also enables the customs administration to share and exchange data with enterprises and conduct online inspections. More advanced ICT applications under development will manage operations involving permit grants, foreign exchange settlements, customs declarations, and tax refunds. These new initiatives aim to make customs administration even easier for firms and further reduce waste and abuse.

The Golden Tax project was launched in 2001; by July of that year, it had connected 31 provinces, more than 300 cities, and over 3,000 counties to the State Administration of Taxation. By 2004, 97 percent of government departments were using online tax registration systems, and 90 percent were using online tax declaration systems and invoice management systems (see figure 6.3). These

Table 6.3 Top 10 Government Web Sites by Type of Sponsor, 2005

Type of sponsor	Sites
Ministerial organizations	Ministry of Commerce, Ministry of Science and Technology, State Environmental Protection Administration, Ministry of Land and Resources, Ministry of Agriculture, Ministry of Foreign Affairs, State Bureau for Production Safety Supervision State Food and Drug Administration Bureau, State Commission of Science Technology and Industry For National Defense, Ministry of Communications
Provinces and municipalities	Shanghai, Beijing, Jilin, Zejiang, Hebei, Anhui, Jiangsu, Yunnan, Shaanxi, Heilongjiang
Cities	Qindao, Wuhan, Hangzhou, Ninbo, Guangzhou, Chengdu, Suzhou, Wuxi, Sanming, Haerbin

Source: CCID 2006.

interconnected systems enable the government to standardize and improve the management of value-added taxes. As a result, revenue from such taxes jumped to $66 billion in 2001—a 17 percent increase over 2000. In 2002, such revenue rose to $76 billion, a 15 percent increase. Since 2003, private enterprises have been required to produce online invoices for value-added taxes.

The Golden Tax project has also helped reduce tax fraud and avoidance. For example, between early 2001 and late 2003 the share of invoices that were forged fell from 0.227 percent to 0.004 percent, while the share of dutiable goods declared by registered businesses rose from 92 percent to nearly 100 percent. The number of large enterprises committing tax offenses has also fallen significantly since the project began. In addition, there have been widespread improvements in government transparency and accountability.

Government Online

China started to push the Government Online project in 2000, encouraging government departments at all levels to establish Web sites. As of June 2005, China had 19,800 domain names and 11,750 Web sites under gov.cn (CNNIC 2005). Increasingly, government Web sites have become a window for government departments to serve the public.

Each year, the State Council Informatization Office ranks government Web sites using a 0–1 scale, based on four subindexes: information openness, government online services, interactivity, and Web design. In 2005, all 76 ministries and their affiliated organizations, 31 provinces and municipalities, 333 cities, and 408 sampled counties were included in this evaluation. Table 6.3 shows the Web sites that received the highest rankings.

On the whole, China's government Web sites remain at the stage of information posting. Information openness has made notable progress at the province and municipality levels, with a subindex above 0.5 (figure 6.4). The most commonly

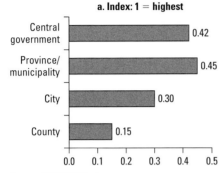

Figure 6.4 Government Web Site Quality, 2005
(Index: 1 = highest)

Source: CCID 2006.

posted information involved policies and regulations, overviews of the area or organization, organizational structures, and contact information.

Although interactive services are the core purpose for government portals, this subindex value was low (below 0.4) at all government levels. More than 80 percent of provincial and city government sites contained citizen surveys and complaint mechanisms, but useful services, such as government feedback and online advice, were rare. Site functionalities—such as content structure, user friendliness, help functions, and quality—also require strengthening, according to the survey results. Maintaining and updating the Web sites remain a challenge (box 6.1).

Box 6.1 A Government Web Site Is Not a One-Off Investment

A 2004 report by the *Economy Reference Newspaper* revealed that more than 40,000 government Web sites in China can be found using search engines. A random sampling of these sites showed that many had only a main page. More than 90 percent of the main pages offered news. However, the information and data were often old; some had not been updated for several years.

For instance, the Web site for the Bureau of Investment Promotion of Ji County (Heilongjiang province) refers to "local government special windows for outside investors." Except for the bureau's address and telephone number, no other useful information is provided. The most recent update was dated January 1, 1970—an obvious mistake.

When a staff member at the Food Bureau of Yan City (Henan province) was asked why the bureau's Web site had not been maintained and updated—the market information for products such as corn, beans, and other products was dated July 2000—he responded that farmers could go to the bureau directly and ask for information they needed.

Source: China Newsnet 2004.

Advancing E-Government

Local E-Government and E-Community

Together with the development of market economy, the society has increasingly higher demand for public services:

- In a dynamic and competitive environment, individuals and businesses demand that the government provide comprehensive and convenient services.

- The disadvantaged in the cities—including low-income households, the unemployed, old persons without family support, and disabled persons—want to get assistance from the government in various areas.

- A large population of migrant workers requires government departments at all levels to ensure effective management and to provide diverse services. Such demand for comprehensive public services has led to the integration of resources and services, enabling innovative local e-government and e-community models.

At very early stage, some e-community applications have grown out of applications initially designed to support administrative functions of the Chinese government (figure 6.5). As a result, many applications described as focused on communities are really aimed at strengthening the government's capacity to provide services to communities. More community-centered services, such as employment information and volunteer service applications, are less widespread.

In recent years, community informatization development has become a priority task for the Chinese government. Different regions have been actively exploring effective models to build harmonious communities and have accumulated preliminary experiences during this process. Some municipal governments have established

Figure 6.5 E-Community Content in China, 2004

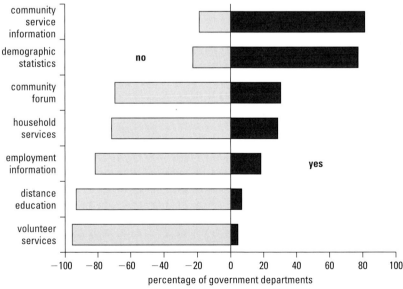

Source: CCID 2005.

integrated e-community service platforms, with call centers and Web sites, in response to increasing demand for information services in large cities. There are e-community applications in central cities such as Beijing, Shanghai, Tianjin, Guangzhou, Wuhan, Hangzhou, Fuzhou, Ningbo, and Qingdao. For example, Yuexiu District in Guangzhou established a one-stop community service system. Residents receive 23 different services at the community centers and are encouraged to provide feedback through Web sites on community planning and administration, thus overseeing government performance. Haishu District, in Ningbo City, has established an interactive community service center accessed using telephone and the Internet. This service platform has helped 12,000 workers find jobs, including 4,000 laid-off workers. Other notable examples at the city level include the community live television Webcast system of Chengdu (Wenjiang District) and the online interview column of focus topics of Xi'an.

Some cities and districts have made distinctive contributions in assisting disadvantaged groups. For instance, Huangpu District in Shanghai set up a social protection "one-stop" service platform, offering services ranging from minimum living guarantee (*dibao*), low-rental housing, and medical assistance, to social relief for original Shanghai residents returning from western provinces. The residents benefit substantially from this effective system. Taking the *dibao* system as an example, the recipients currently spend only three days (15 days before), provide three to nine supporting documents/certificates (five to 19 before), and they do not need to fill out any form (two forms before). Different government departments coordinate internally, and verification of information within a district can be conducted online—previously, the staff had to visit 12 departments to verify information. This helps fill administration loopholes and reflect social fairness. The application and promotion of the *dibao* system in Beijing has also achieved good results. As of July 30, 2005, its information database included 8,190 institutions at city, district, county, village, and community levels, as well as 7,095 registered users. A range of services, such as submission of the *dibao* application, verification, approval, release of funds, reconfirmation of *dibao* coverage, and adjustment of *dibao* standards, are now all computer-enabled. This system handles on average 1,500 cases daily and has become a common social assistance platform for related departments in Beijing.

Rural areas in China could also benefit from online information services. Access to relevant information has the potential to transform economic opportunities and improve livelihoods for rural households. It can do so by facilitating better farming techniques, helping to choose crops to plant in response to market information, reducing exploitation in pricing, creating new possibilities for trade, and improving health and education services.

E-agriculture portals provide three types of information services:

- topics such as weather, crop diseases, and agricultural standards
- government services such as funding and subsidies
- market information for entrepreneurial farmers, such as prices and pesticide standards for exporting to a particular country.

Advancing E-Government

The third service encourages farmers and small agro-industrial businesses to expand their markets and sales.

Although thousands of municipalities in China have e-agriculture portals,[2] many are merely information boards, and much of the information is outdated. Maintaining a resourceful, interactive, and sustainable e-agriculture portal requires considerable resources and commitment from local governments (box 6.2).

Box 6.2 Rural Informatization Case Study of Chongqing

The agriculture sector in Chongqing has gone through intensive modernization since 1999. A management structure for ICT promotion and project implementation within the agricultural sector was established by the local government, following a step-by-step implementation approach.

Some 39 districts and 29 counties have set up information centers within the agriculture department, and the rest have designated one particular office to take on the same responsibility. About 70 percent of organizations affiliated with the agriculture department have established information administration and service departments within their organizations. Finally, 47 percent of townships and villages have established information service centers with both full-time and part-time information officers. Prefecture- and county/town-level local area networks and wide area networks connect all the agriculture and livestock and pastoral departments at these two levels as well as business units under the departments. By 2003, 40 districts and counties had built intranet and e-government platforms.

Chongqing has also constructed a number of shared databases and Web sites that cover topics ranging from agricultural products to technologies and equipment. Market information and price fluctuations are collected and disseminated through terminals in 16 wholesale agricultural markets across the city.

The information service for farmers is provided through three channels:

- The first, farming techniques—inquiry service system, is a call center. Farmers and local farming technicians can call a toll-free number to ask about farming techniques and related information. The callers are answered by an automated system on standard inquiries or receive personal attention from the call center staff. For difficult inquiries, the call center staff members consult experts.

- The second channel, for agriculture network broadcasting, is similar to the first but only provides prerecorded information to local telephone subscribers. This service center is responsible for collecting, editing, and broadcasting relevant information. The program has six sections: daily farming activities, government policies, labor market information, market prices, practical techniques, and inquiries. The monthly subscriber fee is just $0.4 for unlimited access to a specific phone number, and there is no extra charge for telephone connection. After eight months of trial, the system had 26,000 subscribers in October 2005.

(Continued)

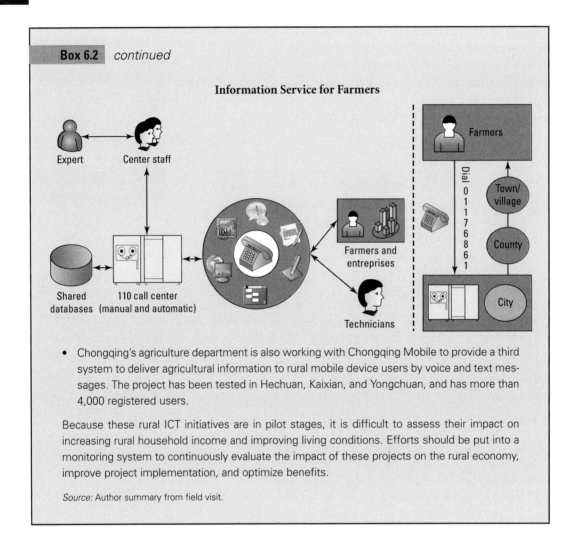

- Chongqing's agriculture department is also working with Chongqing Mobile to provide a third system to deliver agricultural information to rural mobile device users by voice and text messages. The project has been tested in Hechuan, Kaixian, and Yongchuan, and has more than 4,000 registered users.

Because these rural ICT initiatives are in pilot stages, it is difficult to assess their impact on increasing rural household income and improving living conditions. Efforts should be put into a monitoring system to continuously evaluate the impact of these projects on the rural economy, improve project implementation, and optimize benefits.

Source: Author summary from field visit.

Movement Forward

Developing e-government is a massive undertaking with profound effects on the economy, government institutions, enterprises, and individual citizens. The following discussion identifies priorities for China's e-government agenda and obstacles that should be tackled first.

Clarifying the Aims of E-Government through Well-Defined IT Architecture

In developed countries, e-government was initiated as part of reforms to promote transparent governance and open government information in the 1970s. In contrast, many local government officials in China fail to realize that e-government is not about constructing technological networks, but rather about applying technology

Advancing E-Government

for reengineering administrative and business processes, transforming government functions, and improving public services.

China is in the midst of two major transformations: from an overwhelmingly agricultural to an industrial economy and thus from a rural to an urban society, and from a highly centrally planned economy to a market economy. In this context, a major challenge is using ICT to help make the government more effective, efficient, and transparent, to facilitate these transformations. ICT is a means, not an end in itself.

Some of the more successful countries have adopted enterprise architecture as an approach to e-government. For example, Canada has established the Business Transformation Enablement Program, which seeks to integrate business processes across the government to provide seamless public services. The United States has created a Federal Enterprise Architecture for ensuring the development of e-government applications in a coordinated manner.

It will be important for China to embrace IT architecture for the successful deployment of e-government services. When defining its IT architecture, the Chinese government needs to revisit organizational structures for guiding ICT development. This will be vital to the coordinated and coherent development of e-government applications across different wings of the government.

Such an approach to e-government can help China's central government coordinate government departments and institutions at different levels, planning both vertically and horizontally and integrating government information resources to connect information "islands" at all government levels. As the development of government-to-business (G2B) and government-to-citizen (G2C) applications accelerates and the quality of service improves, businesses and citizens will experience more transparent, responsive, and efficient government services, and subsequently increase their support for e-government. Thus the openness of information flows to the public must be improved; the continuous process of generating, updating, absorbing, and exchanging information resources as part of a well-defined IT architecture remains key to achieving its huge economic and social benefits.

Using Public-Private Partnerships

For a long time, China's e-government applications have been led by technology rather than driven by demand—leading to a structural imbalance between supply and demand. ICT vendors tend to build networks, databases, hardware, and software based on existing technologies and products. Many do not have the capacity to analyze the types of services that are in demand and, as a result, cannot provide mature solutions to the problems the government urgently needs to resolve. Hence the supply and demand of e-government applications need to be carefully matched.

The use of public-private partnerships (PPPs) can accomplish the following:

- assist in leveraging private sector funding for delivery of e-services
- give the government access to advanced ICT project-management skills
- accelerate e-government project implementation

- help the private sector better understand demand and offer a solid and continuous market base.

While PPPs are not a panacea for all the challenges involved in implementing e-government, they can provide distinct value for a range of e-government applications and offer relevant, cost-effective solutions.

China has made limited use of PPPs in e-government. In response to rapid economic and social development, the government's functions are shifting from micromanagement to macro-coordination. Its agenda includes large initiatives such as developing a social security system, providing employment for agricultural workers, maintaining economic growth, and reforming and building the tax system. The challenge will be to entrust the private sector to participate in such initiatives while creating incentives for the country's application providers to develop the ability to support such complex tasks.

As encouragement for greater adoption of PPPs, it will be necessary to develop a framework for e-government PPPs at the central, provincial, and local levels, and train key government staff members on the planning, design, development, implementation, and monitoring of PPP projects.

Enhancing Quality Standards

The link between the government and ICT suppliers is only part of the story. The ultimate beneficiaries of e-government projects are Chinese citizens and businesses. The quality and standards of ICT application providers must improve to meet the demands of these users. The government has experimented with the use of third-party supervision to ensure the quality of e-government projects. For example, companies like Qing Hua Wang Bo (an IT service provider linked to Qing Hua University), Ken Si Jie (a computer system research institute), and China Software Testing Center have been involved as auditors of government projects. It would be desirable to create capabilities for defining and monitoring standards for application and process design in e-government. Leading universities and quality standards organizations would be natural partners for such an initiative.

Integrating and Sharing Information Resources

Gaps in economic development have resulted in huge differences among Chinese regions in terms of information resource development and use, giving rise to a digital divide in addition to those for information infrastructure and technology (see chapter 3). The annual survey of government Web sites by the State Council Informatization Office shows that the top 10 local government sites are all in the eastern region (see table 6.3). In particular, Beijing, Shanghai, Jiangsu, Shandong, and Guangdong have a large lead in e-government applications over other provinces. As for e-communities, few demonstration projects are from the central and western regions. Among the 400 million citizens covered by the national population information database, 70 percent are from the eastern region.

Advancing E-Government

There is a clear need to establish cross-regional mechanisms for sharing and exchanging information resources. To a large extent, development in China's eastern region depends on the huge pool of low-cost human resources and raw materials from the central and western regions and would benefit from those regions achieving sustainable development.

The government needs to look at some of the emerging models for sharing and collaborating on e-government applications from other parts of the world. For example, the Association of Developers and Users of Open Source Software in Administrations and Local Communities (ADULLACT) in France has members from local authorities, associations, and service companies. ADULLACT has emerged as a platform for collaboration and sharing of applications by local authorities, thus avoiding duplication and reducing the time and cost of developing e-government solutions. Other initiatives include Ireland's Local Government Computer Services Board and the United Kingdom's Local Authority Software Consortium. While these examples relate to local governments, the logic would apply equally to provincial, national, and regional government entities. For example, the European Union has set up an e-Government Observatory to share best practices and help in the interoperable delivery of European e-government services to public administrations, businesses, and citizens.

China should consider creating an institutional mechanism for integrating and sharing e-government applications across provinces and local governments. This could prove especially helpful to those regions and provinces currently lagging on e-government application development.

Maximizing Investment Returns on E-Government Projects

E-government applications are often huge management information system projects that require large investments. At the same time, one of the main objectives of e-government is to reduce the transaction costs of government operations. In China, many investments in large-scale informatization projects have had a mixed record, with some projects exceeding expectations while others have yielded limited results.

A lot of this investment went into telecommunications infrastructure, networks, and databases. While these provide the foundation for ICT applications, such investment was sunk and construction is often repeated, resulting in overinvestment and overcapacity. At the same time, returns are small when e-government applications are at an early stage of development. In addition, there have been more government automation projects (including government-to-government, or G2G, applications) than government-to-citizen or -business services thus far, so the general public may not be aware of the savings resulting from G2G—but have high and unmet expectations for G2C and G2B services. For example, only 3 percent of Internet users accessed e-government applications in 2005, and only 11 percent were satisfied with the e-government information available (CNNIC 2005). Moreover, impacts of ICT applications are hard to quantify. All this leads to reduced support and participation in e-government initiatives.

It would be useful to adopt a clear methodology for prioritizing e-government investments. For example, Australia has established a demand-and-value assessment methodology for objectively determining which e-government applications deserve funding. The United States has a similarly well-structured process as part of its Performance Reference Model. The Chinese government would be well advised to consider such an approach so that investments in e-government can yield tangible results. This apart, the importance of monitoring and evaluation frameworks with clear and measurable output and outcome indicators for each project cannot be overemphasized.

Making Information Flows More Transparent

The primary goal of any e-government application is to achieve transparent information flows. Like many countries, China has a long, entrenched tradition of government secrecy. Chinese scholars estimate that some 80 percent of useful information in China languishes in government files (Horsley 2004). This culture of secrecy has meant that the government acts as a bottleneck to the free flow of economic, social, and other information that would facilitate continued dynamic growth and development. Officials and scholars alike have noted that the lack of transparency contributes to corruption, misallocation of resources, and distrust of public institutions.

Information access is likely to improve in the near future. Provisions on open government information have been drafted (China Newsnet 2004). The draft provisions establish a presumption of disclosure, making secrecy the exception rather than the rule. They provide that citizens, legal persons, and other organizations have the right to request government information from government agencies, including information about individuals themselves, and refer to the right to know. The provisions impose a legal obligation on government agencies to disclose all information not covered by a specified exemption.[3] In addition, such information must be posted on the government's Web site.

It would be interesting to emphasize the performance of different government departments and agencies in achieving transparency (within the information openness subcategory) in annual e-government Web site rankings and to give the matter wide publicity so as to motivate government entities to be more transparent.

Notes

1. See http://tech.sina.com.cn/i/w/2004-01-13/1043281578.shtml.
2. The national agro-industry's Web site provides daily postings of price information for more than 300 products in 280 large, wholesale agricultural markets. This e-government service also publishes over 300 pieces of information a day for more than 25,000 customers. The site is visited by over 15,000 users a day, and about 9,000 village service points have been linked to this system (World Bank 2006a).
3. All government information is to be accessible by the public unless it falls within one of six listed exemptions from disclosure: a state secret; a commercial secret; an individual's private information; information related to a matter being investigated, discussed, or processed; information related to an administrative enforcement action that might influence the enforcement activity or endanger an individual's life or safety; or information otherwise exempted from disclosure by law or regulation. All but the state secrets exemption may be subject to a balancing test.

Chapter 7

Fostering E-Business

Competition and globalization will create further demand for informatization to increase domestic firms' productivity and efficiency. As China begins fulfilling the terms of its accession to the World Trade Organization (WTO) by further opening up its economy, firms will need to become more competitive to survive and grow in domestic and foreign markets.

Informatization at the enterprise level (defined as e-business in this publication) broadly falls into two categories:

- *Internal informatization applications, providing firms with opportunities to enhance productivity and efficiency.* These mainly consist of increasing reliance on information and communication technology (ICT) for basic management and operational functions (such as office automation and management information systems). Applications of ICT within and across the structures of firms (such as encouraging e-mail communications internally and externally, including through Web sites) are included, relying on standard electronic procedures and business processes and allowing efficient cooperation with buyers, sellers, and partners. Recent developments in business process outsourcing have increased interest for such applications across Chinese firms.

- *E-commerce applications, allowing firms to carry out contractual transactions with other businesses as well as with individual consumers, business to business (B2B) and business to consumer (B2C), typically over the Internet.* E-commerce will allow Chinese firms to become part of complex integrated networks of players involving various online business processes (such as supply chains) and to contract, buy, and sell online. The growing importance of e-government also creates new opportunities to conduct business-to-government (B2G) transactions in the context of public procurement, for example, but also through public-private partnerships in the delivery of public services to citizens.

China's Information Revolution

Figure 7.1 Informatization at Enterprises

Push/pull factors	Consequences for business	Effect on informatization
Increased competition (e.g., globalization of markets, WTO, and lower margins)	Need to streamline and modernize business processes and management techniques	More internal informatization
Integration of supply- and delivery-chains, inventory management, etc.	Need to adopt electronic procedures compatible with those of suppliers, clients, and partners	
Emergence of new opportunities for business (online buying and selling)	Will have to exploit online opportunities, have a Web presence, reach new markets, and disintermediate transactions	Growth of e-commerce

Source: Author.

Internal applications are not necessarily a prerequisite to participate in e-commerce. For example, firms with limited internal informatization capacity can use e-commerce platforms offered by a third party (private or public) to conduct online sales and purchases. More often than not, however, automating internal operations is the first step for firms that want to pursue market expansion and increased profits through Internet sales (figure 7.1).

Chinese firms are relatively aware of the positive effect that informatization can have on their operations. In that context, their two biggest goals for informatization are to improve efficiency (such as in supply chain management) and reduce sales costs by automating transactions and conducting them online (figure 7.2). Other objectives indicate that ICT use is viewed as a way to strengthen links with suppliers

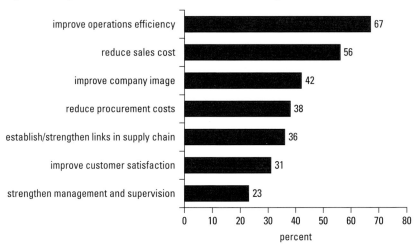

Figure 7.2 Objectives of Informatization for Chinese Enterprises, 2003

- improve operations efficiency: 67
- reduce sales cost: 56
- improve company image: 42
- reduce procurement costs: 38
- establish/strengthen links in supply chain: 36
- improve customer satisfaction: 31
- strengthen management and supervision: 23

Source: Ministry of Commerce 2004a.
Note: Multiple answers were allowed.

Fostering E-Business

Figure 7.3 Chinese Firms' Investments in Informatization by Industry, 2003 ($ *thousands*)

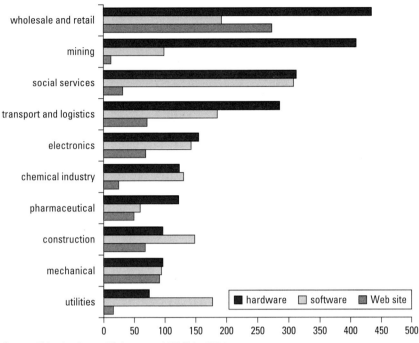

Source: China Academy of Sciences and HP China 2004.

and customers and to improve management capacity. These are all important elements for firms to stay competitive in rapidly evolving markets.

Chinese firms' investments in informatization have grown steadily in recent years. A 2004 survey by the Ministry of Commerce looked at 838 firms with annual sales above $400 million and found that 43 percent increased such investments by about 20 percent between 2002 and 2003. Furthermore, 38 percent of the firms maintained their ICT investment levels, and only 7 percent reduced them (Ministry of Commerce 2004a). Among industries, wholesale and retail firms invested the most in ICT hardware, with an annual average of $430,000 per firm, followed by mining firms at just over $400,000 (figure 7.3).

Wholesale and retail firms also spent the most on Web site construction and maintenance, averaging $260,000. For software, the social services industry invested the most, followed by wholesale and retail and transport and logistics (China Academy of Sciences and HP China 2004).

Different ICT investment levels across industries show that, beyond a certain level of connectivity (computer use, Internet access, online information, and marketing), some firms in China may choose to stay with traditional business processes either because more sophisticated ICT applications are unsuitable for their sector or because expected returns are too small. Any supporting role played by ICT is crucially dependent on the business processes required by a firm and its broader industry environment (Qiang, Clarke, and Halewood 2006). Thus, it is important for a firm to decide whether and how much informatization is necessary for that firm's particular operations, and what kind of ICT applications are best suited to help improve performance.

Internal Informatization

For many firms the initial objective for adopting ICT applications is to integrate fragmented and departmentalized internal information and processes. Office automation (OA) and management information systems (MIS), which transfer information through shared electronic files and networked computers, have the potential to improve the efficiency of business processes such as documentation, data processing, and other back-office functions (Qiang, Clarke, and Halewood 2006).

As Chinese firms evolve, they require increasingly sophisticated ICT applications, such as business process management (BPM), customer resource management (CRM), enterprise resource planning (ERP), and materials requirements planning (MRP). These applications allow firms to store, share, and use information seamlessly and reduce inefficiency and operational and transaction costs, thereby improving productivity and profitability (box 7.1).

Box 7.1 **Examples of Enterprises' Internal Informatization Applications**

Office automation (OA) is the use of computer systems to execute a range of office operations, such as word processing and email. OA almost always implies a network of computers with a variety of available programs and is the first step in creating a shared access filing environment.

A management information system (MIS) goes a step beyond automating a single function by organizing and linking a group of interdependent items. Although such systems—sometimes referred to as JinXiaoCun (or JXC, with Jin meaning supply, Xiao meaning sales, and Cun meaning inventory)—are fairly prevalent among high-technology manufacturers in China, they provide limited functionality. For example, some do not integrate finance, procurement, sales, and inventory parts of the business over the Internet. Materials requirements planning (MRP), narrowly defined as a type of MIS (or JXC in the Chinese context), is another basic information system used to provide supply, sales, and inventory support.

Chinese firms use enterprise resource planning (ERP) systems for financial services, distribution and sales, billing, shipping, material consumption, and production planning. Firms implement ERP in one of three ways: by concentrating finance and accounting at the core of the system; by concentrating MRP, inventory management control, and production planning at the core; or by concentrating order management and customer service at the core and thereby tracking sales.

Large firms and banks use customer resource management (CRM) systems to analyze the profitability of businesses, and companies use them to process large customer volumes and determine business strategies based on the profitability of high-value customers. The CRM systems of most Chinese high-tech manufacturers still involve significant manual processes, and most companies have no online archives offering data on historical relationships.

Source: Author analysis.

Fostering E-Business

Most medium-size and large Chinese firms have automated one or more of their office operations. Computer applications were introduced mainly by large firms in the early 1990s, often to automate accounting processes. Since then, the use of ICT-based accounting applications has grown considerably: in 2004, 98 percent of large firms and 90 percent of medium-size firms used them. The key growth strategy for large firms has been the application and integration of information systems to streamline firm processes such as supply chain management and product data management. These large firms include state-owned firms, of which 22 percent have integrated OA systems.

The manufacturing sector is China's largest and covers many industries. During the 9th and 10th Five-Year Plan periods (1996–2005) the sector drove enterprise informatization and received government funding and support for such efforts. As a result, about 90 percent of firms in the automotive and electronics manufacturing industries now have Internet access and use ERP applications, and nearly all use computer-aided design (CAD) software (figure 7.4). Internet access and ICT applications are also widespread among manufacturers of electric appliances and machines (an industry that is a pillar of the manufacturing sector), and garments.

ICT applications have significantly improved production and management systems in China's industrial firms and play an increasingly important role in supporting their growth. Yet a 2003 survey of Chinese firms conducted by the Ministry of Commerce found that many did not have any internal informatization applications

Figure 7.4 Internet Access and ICT Application Use in China's Manufacturing Industry, 2003

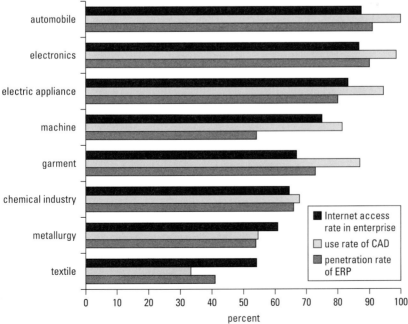

Source: CCID 2005.

Figure 7.5 Prevalence and Reported Impact of ICT Applications in Chinese Firms, 2003

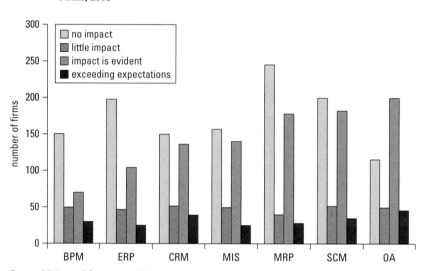

Source: Ministry of Commerce 2004a.

(figure 7.5). Such firms may not have found an ICT application suitable for their operations. Among firms that had adopted specific ICT applications, most found that the impact on their operations met or exceeded expectations.

Chinese Firms Using E-Commerce

A significant proportion of Chinese firms has started to move beyond informatization of internal information and processes to enter the arena of e-commerce. Models such as B2B (sometimes B2G when the buyer is government) and B2C have started to emerge as profitable activities. E-commerce in China took off in 1999 as dot-coms such as sohu.com, sina.com, China.com, and Netease (www.163.com) went public on the NASDAQ stock exchange.

The Chinese Internet sector rapidly attracted significant investment flows. E-commerce revenue totaled $2.2 billion in 2000, with B2B accounting for $1.7 billion and B2C for $0.5 billion. Although e-commerce slowed during the crash of 2000–01, it recovered when the three largest portals (Netease, sina.com, and sohu.com) began to make profits in 2003 (Ministry of Commerce 2004a). In 2005, China's e-commerce market grew by 42 percent to $84.2 billion (CCID 2005).

Business-to-Business E-Commerce

In 2005, B2B revenues in China totaled $80 billion, or 95 percent of the e-commerce market. In 2004, China was home to more than 4,000 e-commerce Web sites, mostly supporting B2B activities. This number is increasing more than 10 percent a year.

Figure 7.6 Changes in Supplier and Client Contacts among Chinese Firms Engaged in E-Commerce, 2003

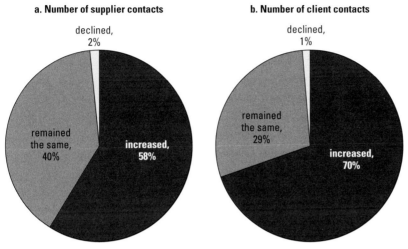

Source: Ministry of Commerce 2004a.

These sites help businesses link to suppliers and clients, providing opportunities to improve product supply chains and reach potential clients. The 2004 Ministry of Commerce survey of 838 firms found that 58 percent of those that had participated in e-commerce increased their number of supplier contacts, while 70 percent increased their number of client contacts (figure 7.6).

A few large Chinese companies in the automobile, iron and steel, and electronics industries have led the way in developing B2B e-commerce and incorporating ICT into their purchasing, advertising, and marketing. B2B e-commerce has become a key element in improving the efficiency of supply chain management of many large Chinese firms (box 7.2).

B2B e-commerce systems often rely on sophisticated Internet tools and applications, because they require large amounts of data and commodity exchanges, and subsequently require large Web capacity and back-office OA systems and MISs to manage the inflow and outflow of information. Firms in China, especially small and medium-size enterprises (SMEs), still find it difficult to participate in B2B e-commerce. The reasons mainly have to do with their lack of expertise in ICT applications and lack of knowledge in dealing with new modes of online transactions and e-contracts (figure 7.7). Although in theory online transactions and e-contracts should be as valid as their traditional, paper-based counterparts, firms still have to make the transition from brick and mortar operations to the online environment. SMEs have difficulty doing so, particularly in finding the funds and information technology (IT) expertise to integrate their internal operations seamlessly with online business transactions and functions.

In response to these constraints, off-the-shelf B2B platforms have become popular, providing an intermediary where businesses can join and search for potential

> **Box 7.2 Large Firms Lead in B2B E-Commerce**
>
> The following large firms are leaders in B2B e-commerce in China today:
>
> - *Automotive Works Corporation and Dongfeng Automobile Group,* two giants in China's automobile industry, jointly established an e-commerce procurement and distribution platform in 2000. The platform integrates 10,000 automotive spare parts and components firms and involves annual transactions of more than $1.5 billion.
> - *Shanghai General Motors* has actively promoted flexible production and has restructured its management structures and business operations to optimize its supply chain. It uses a network platform to communicate and coordinate with more than 200 spare parts and components suppliers in China.
> - *Baoshan Iron and Steel* has established an efficient, responsive, and integrated supply chain for its customers, suppliers, and employees. Through its B2B e-commerce platform, Baosteel, the company has purchased $2.4 billion in goods from some 500 suppliers. Each year the company sells iron and steel products worth billions of dollars, and a large share of its export transactions are conducted online.
> - *Haier Group,* a major maker of consumer electronics and appliances, conducts some procurement online, offering tenders to a network of some 400 suppliers—significantly shortening the time required for procurement, distribution, and logistical functions.
> - *Lenovo,* a computer manufacturer, has an e-transaction system linking agents, distributors, and vendors with its internal application systems. Lenovo now procures online some 20 percent of what it requires for its supply chain.
> - *PetroChina's* B2B e-commerce system (www.chinaoilweb.com) is widely used, covering 9 provincial branches, 260 prefecture branches, and 1,400 county branches. PetroChina has also continuously improved its supplier management systems. Through its Web site, www.energyahead.com, the company has integrated almost 2,500 suppliers of nearly 80,000 products in dozens of categories.
>
> *Source:* Author analysis.

clients and suppliers in a sort of commercial information exchange. Common platforms provide a global base that enables firms to search for business opportunities in parts of the world where they otherwise would not have the necessary contacts. They also allow smaller firms to access commercial information and bid for business projects and deals with larger, established firms. Firms can also lower costs by conducting logistical and operational transactions online. Alibaba China, for example, is the largest online marketplace for domestic SMEs to advertise their products and find trade opportunities (box 7.3).

Fostering E-Business

Figure 7.7 Main Obstacles to E-Commerce in China, 2003

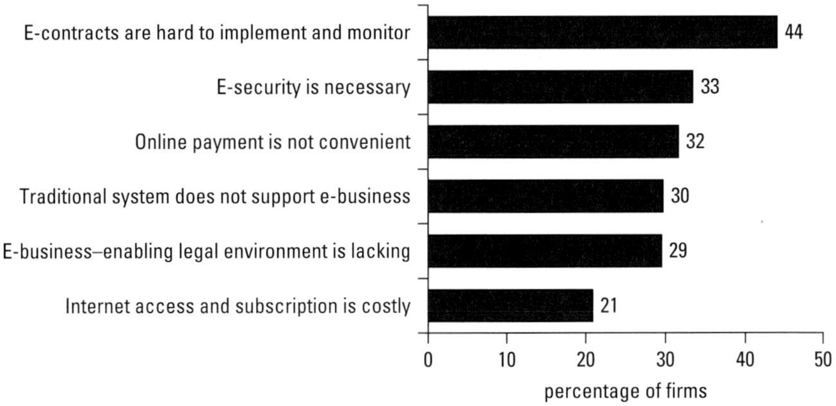

Obstacle	Percentage of firms
E-contracts are hard to implement and monitor	44
E-security is necessary	33
Online payment is not convenient	32
Traditional system does not support e-business	30
E-business–enabling legal environment is lacking	29
Internet access and subscription is costly	21

Source: Ministry of Commerce 2004a.

Box 7.3 Alibaba China

Alibaba.com Corporation (Alibaba China) was founded in 1999 in Hangzhou. Alibaba China operates China's largest online marketplace for domestic B2B trade, as well as China's most popular online payment system, AliPay. It has 7.1 million registered members, mostly SMEs from mainland China, Hong Kong (China), Macau, and Taiwan (China) who pay an annual subscription fee that entitles them to post trade offers and products online. In addition, more than 100,000 businesses pay $250 to $10,000 a year for Alibaba China's online marketing services. Alibaba also receives commissions for transactions conducted through its platform.

The information provided on Alibaba's Web site includes company profiles and product information and exhibition. Industry news and trends are tailored to each member. There is also a feedback service through which members can contact each other and make price inquiries, conduct negotiations, and place orders. The correspondence is managed and filed by Alibaba China, ensuring data security. Other services provided on the Web site include value-added services provided by third parties in such areas as advertising and marketing consulting, authentication and verification of a member's identity, and logistics services.

In October 2005 Alibaba.com acquired Yahoo! China, whose search properties combine globally and locally developed technologies to provide relevant Chinese-language results. It will play a valuable role in powering Alibaba.com's vision to become the top destination for Chinese businesses to find trade opportunities, promote their products, and conduct transactions online.

Source: http://China.alibaba.com.

B2B e-commerce platforms based on a specific industry or niche market have great potential. For example, the China Chemical Network (www.chinacheminfo.com) is an influential Web site to facilitate international trade in the chemical industry by providing exporting domestic firms with information such as business regulations of other countries and price quotations.

Business-to-Consumer E-Commerce

B2C e-commerce is a key aspect of informatization for Chinese firms that sell their end products to consumers. Until the Internet bubble burst in 2001, many B2C Web sites had been operating at a loss, focusing on attracting users rather than generating revenue. B2C has been growing since 2001; in 2005, B2C revenue totaled about $4 billion.

Many large and established firms in China offer a wide range of products online. They carry out the entire operation on their own, from setting up their B2C Web sites to marketing and distributing their products to online consumers. They may already have a working logistics and distribution system across China. This is an advantage if they are able to integrate logistics IT applications seamlessly with the new B2C platform and to train managers to adapt to the new system and perhaps handle increased product distribution.[1]

SMEs, however, often rely on intermediaries to provide their B2C platforms. Examples of B2C platform providers include Sina and eNet. Such platforms target small businesses or traditionally offline companies that want to conduct online sales but may not have the capacity to create and maintain their own B2C platforms. These B2C platforms use third-party delivery channels such as China Post, EMS, and Eastern Express to deliver products to consumers.

Although larger players in the market may benefit from their own B2C platforms and logistics and distribution channels, margins can be limited because of high operating costs and intense price competition. In certain cases, firms that are doing exceptionally well are those that have focused on a market niche and limited distribution area. For example, joyo.com, the largest online seller of books, music, and videos in China, limits its distribution destinations to urban Beijing, Shanghai, and Guangzhou. My8488 E-commerce Co., another successful example, limits where products come from, selling only products available in wholesale markets in the Linyi area.

China is a large country, with an enormous population, wide territory, and large imbalances in regional economic development and gaps in urban-rural incomes. For the foreseeable future, Internet users will remain concentrated in large and medium-size cities and relatively developed coastal areas. Localizing a firm's customer base, resource planning, distribution system, and market promotion can help create a critical mass of online consumers and realize economies of scale.

Although B2C e-commerce in China has experienced considerable growth in revenue, the number of online shoppers is still relatively small. A July 2005 survey by the China Internet Network Information Center (CNNIC) found that only 7 percent of the country's 111 million Internet users made purchases online—much less than used other online services such as e-mail (91 percent), news retrieval (79 percent), Web searches (65 percent), or online music (46 percent).

Among consumers who did shop online, more than half had made just one to three purchases and spent less than $60 in the past six months (figure 7.8). Online shoppers acknowledged that online purchasing was easy and offered a large variety of goods at low prices—an indication that Chinese Internet users have accepted online buying as a commercial channel. However, after-sale services, e-security, and

Fostering E-Business

Figure 7.8 Frequency and Spending of Online Shoppers in China, 2005

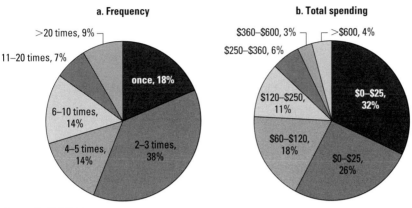

Source: CNNIC 2005.
Note: Data refer to the six months before the survey.

Figure 7.9 Main Drawbacks to Online Purchases in China, 2005

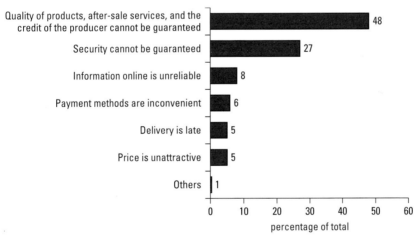

Source: CNNIC 2005.

the uncertain reliability of producers and online information were cited as major drawbacks to online shopping (figure 7.9).

China's online payment system is improving, especially on several large e-commerce platforms. For instance, the Capital E-Shop platform supports online payments by 60 bank cards issued by 10 domestic banks and 4 international credit cards. In addition, the proportion of Internet shoppers who found making payments online inconvenient dropped from 17 percent in 2000 to 6 percent in 2005 (CNNIC 2005). About half preferred online payments (credit or debit cards) to other methods (such as payment on delivery, bank transfer, or postal order), up from 25 percent in 2000.

Still, there is much room for improvement in the services that accompany online purchases, such as after-sale services and return and credit policies (see figure 7.9). Furthermore, as e-commerce matures, the market for Internet user profiles will grow, enabling advertisers and marketing firms to tailor their promotional programs to targeted audiences. Consumers' data and privacy need to be protected as they surf the Web.

Movement Forward

The forces that encouraged enterprise informatization in China will still be a vital part of the Chinese and international economic environment in coming years. In addition to the trends identified and analyzed in this chapter, new avenues have started to emerge as promising areas for Chinese enterprises, including m-commerce (transactions conducted using mobile telephones) and B2G (on-line transactions between businesses and governments). New business models are also affecting many online services, as Internet-based advertising has started to grow faster than other forms of advertising (including through newspapers and broadcasting). Hence, e-business will continue to grow rapidly in China.

In this highly dynamic environment, the Chinese government has a large array of responsibilities to fulfill in promoting and accompanying enterprise informatization. Key action items are analyzed in the following sections.

Encouraging Enterprise Informatization through E-Government Initiatives

The government can encourage traditional brick and mortar firms to engage in e-business by strengthening its role as a supportive administrator through national and local e-government initiatives and programs. For instance, the Board of Supervisors for Major State-Owned Enterprises, which oversees the 100 largest state firms, encourages them to adopt IT and e-commerce to cut costs and increase productivity, among its other mandates. Program 863 (started in 1986) launched about 190 ICT projects in 2002, resulting in an increase in the use of ICT applications and the advancement of management concepts in the manufacturing, retail, energy, telecommunications, construction, and textile industries. The program has supported more than 2,000 firms in 27 provinces.

In addition, government provision of online information and services can demonstrate the effects of ICT on businesses by spreading awareness of the potential for online delivery and interaction, and help build trust and security in online transactions (see chapter 6).

As model users, governments (central and local) can also act as standard-setters for ICT adoption by firms. To ensure access to public services and meet obligatory requirements for business purposes, firms would be encouraged to adopt ICT and adjust their choices of systems and software to maintain interoperability with

Fostering E-Business

existing online government services. Public e-procurement provides such an example (Qiang, Clarke, and Halewood 2006).

Improving the E-Commerce Environment

Regulatory measures will need to be taken to help standardize the e-commerce marketplace, lower barriers inhibiting e-commerce activity, and foster market growth. Building on the E-Signature Law,[2] the government's further support for a regulatory framework that builds trust and security and combats cyber crime is essential in encouraging business use of ICT applications.

Encryption technologies must be made more widely available to enable secure online transactions. Although the E-Signature Law did much to support e-commerce and online financial transactions, authentication mechanisms and antifraud measures—combined with privacy and consumer protection protocols—are crucial for more Chinese to feel safe shopping online. Of particular relevance for firms are low-cost online dispute-resolution mechanisms, both among firms and between firms and consumers (Qiang, Clarke, and Halewood 2006). The government may also want to consider a light touch when taxing online transactions, to avoid stifling growth in a fledgling area.

Establishing National Logistics and Distribution Systems using Advanced Information Technologies

The promotion of e-commerce in China has increased requirements for the development of logistics and distribution systems, as well as offered significant new opportunities. Because logistics in e-commerce have characteristics of widely distributed consumers—small volume, multiple batch, and short cycle—traditional logistics have had to make significant progress on reaction speed, stock control, and service quality.

Logistics and distribution systems should adopt informatization themselves, using advanced technologies such as barcode readers, electronic ordering systems, package tracking systems, effective customer reaction, and global positioning systems to integrate traditional distribution and financial systems with modern information networks. The goal is to efficiently coordinate the transportation network, find the nearest goods resources for each order, reduce logistics expenses through accurate information, and arrange the best transportation strategy as a whole. Such efforts may include modernization of the postal system, as well as the opening of specific postal and delivery services to competition.

Incubating SMEs and Supporting Research and Development

According to the Asia-Pacific Economic Cooperation forum, there are 8 million SMEs in China. Most are still in the early stages of ICT application deployment—possibly because many do not have the means to access such applications or the expertise to implement them. Firms that want to pursue e-commerce but lack the means to do so can be assisted through government-sponsored B2B and B2C platforms. Such initiatives could be coupled with promotional programs and e-commerce trade shows that demonstrate how e-commerce works and the necessary steps for online transactions. The promotional programs could also encourage SMEs that do not currently export to do so through e-commerce.

Research and Development is increasingly critical to competitiveness, as is an understanding of market needs. The Chinese government needs to provide incentives and support for Research and Development in application development (see chapter 4). It can also assist firms by putting in place complementary institutions and associations that provide advice on technology choice, identify and disseminate information on best practices, and assess e-commerce technologies and approaches (Qiang, Clarke, and Halewood 2006).

Strengthening International Competition and Cooperation Online

The government should continue to encourage Chinese companies to purchase and sell goods and services from and to both domestic and foreign businesses operating online, to reap the rewards of reaching suppliers and consumers worldwide. In the near future, China will likely lead the region (and the world) in the volume of e-commerce transactions and number of e-commerce trade partners. Thus China could play a larger role in establishing and furthering regional and international e-commerce standards. A harmonized legal and regulatory framework would benefit both individual countries and the region as a whole, instilling trust in this new mode of trade and cooperation.

Notes

1. Distribution channels for B2C e-commerce are different in that there are many more destinations for products sold online. The distribution center can be considered the last link between the product and the customer, so the management and operation of the distribution become critical when dealing with individual orders and addresses. Delays in delivering products and mistakes in orders will create customer dissatisfaction, which is an important determinant for online shoppers.
2. In recent years the Chinese government has taken significant steps to improve the regulatory environment for informatization, including laws and regulations at the firm level (see chapter 2). The E-Signature Law was an important milestone (see box 2.1). To help implement the law, the Ministry of Information Industry issued administrative measures on electronic authentication. Both the law and the administrative measures took effect in April 2005.

Chapter 8

Connecting the Issues: A Summing Up

Despite significant progress in the past decade, obstacles remain to accelerating informatization in China. To a large extent, these are systematic problems cutting across the economy and society, tied to China's ongoing economic and social transformation. Addressing all the critical factors is complex and requires long-term commitment. To trigger broader changes, several key issues need to be addressed decisively in the second half of this decade through policies entailing institutional reform. This section draws together these issues and examines their strategic implications.

Promoting Indigenous Innovation for Domestic and Foreign Markets

China aims to become a center of innovation, content creation, and research and development (R&D). Among the dilemmas Chinese policymakers face is whether to focus on low-cost solutions to domestic challenges (such as equipment and services affordable for the mass market) or to focus on competitive alternatives to imported information goods and services—while competing successfully in global markets for such products and services.

With respect to informatization, those two goals are not necessarily conflicting. For example, China could provide incentives to local computer manufacturers to develop low-cost personal computers to increase penetration among lower-income households. The large Chinese market means that producers can achieve significant economies of scale with consequently lower prices. This could have tremendous export potential because many developing countries face the same challenge of being unable to afford information and communication technology (ICT) products.

At the same time, China should move away from purely labor competitiveness in the export market. Domestic ICT industries continue to rely significantly on overseas technologies. Domestic industrialization and upgraded consumption, especially in urban areas, require China's ICT products and services to move up the value chain. Otherwise, the paradoxical situation would continue where China exports 90 percent of locally manufactured chips, then re-imports a large share after foreign companies process them for Chinese consumption.

To make the best use of its resources and competencies, China should focus R&D efforts on core technologies, products, and services that have been identified as important for informatization, that necessitate extensive customization or are unique to the Chinese market, or where it has strategic advantages. These include items such as integrated circuits, network security software, telecommunications equipment, and mobile data applications.

The most critical role for the government is at the initial stage of innovation, where the eventual success of a technological invention is highly uncertain and potential investors are too risk-averse to finance its development. The state could assume part of the risk or help in facilitating movement through the innovation cycle.

Creating Strategic Engagement in Standards

The Chinese government might be tempted to promote domestic standards for several reasons. First, China is caught in a technology trap. Lacking advanced technology of its own, its exporters have to license much of that technology from abroad, often at high cost. Giving Chinese firms a head start by creating home-grown standards and know-how, and using the domestic market as a springboard, offers a way out.

Second, national prestige is at stake. Control over technical standards can determine the rise or fall of domestic industrial empires. As Chinese officials like to say, third-class companies make products, second-class companies develop technology, and first-class companies set standards (Suttmeier, Yao, and Tan 2006). In the case of broadband infrastructure, for instance, China plans to take the third approach. It is backing its ambition to develop from a basic manufacturing economy into an advanced industrial power with heavy state spending on research and showcase projects to promote indigenous innovation.

For all of China's outward reserve, it is unclear whether such industrial policies will pay off. The ICT market is very dynamic, moves quickly, and is internationally integrated. China should be cautious about too strong an involvement in driving the development of technologies inconsistent with global standards. Unless other countries adopt its standards, pursuit of *techno-nationalism* may lead to isolation rather than global supremacy. The collapse in world market share of Japan's mobile telecommunications manufacturers since it embraced its own unique standard in the early 1990s is a cautionary lesson (de Jonquieres 2006). Moreover, developing domestic standards could make it harder for Chinese firms to be successful in foreign markets if they have to develop different export products based on global standards and could discourage international information technology (IT) firms from investing in China.

Connecting the Issues: A Summing Up

It will require patience and maturity to continue to work in and through international standards bodies to build the complex coalitions required, receive recognition of Chinese technologies, and resist the temptation to follow independent routes. When it comes to standards, strategic engagement is likely to be more successful than isolationism.

Thus, an effective approach for China would be to use its large and growing market to influence global standards in ways favorable to its firms. For instance, it could encourage the adoption of open rather than proprietary standards to reduce royalty burdens on its companies and allow them to develop technological expertise within open-technology platforms. It could also promote the use of innovative IT applications that fit domestic needs to encourage technology vendors to use China as a test market for new standards.

Matching Supply and Demand of Commercially Successful Applications

Although Chinese science is developing rapidly, as reflected in growing numbers of patent filings, the country's efforts to translate ideas into commercially successful innovations have so far been disappointing. Many structural barriers stand in the way. They include an ivory-tower approach to engineering education, weak links between universities and industries, ineffective intellectual property protection, state-owned industries' domination of large markets, and scarce venture capital funding.

China faces the paradoxical situation of being a leading exporter of high-technology equipment while being unable to meet its own ICT needs. Many of its ICT vendors do not have the capacity to analyze the types of services that are in demand; as a result, these vendors cannot provide mature solutions to the challenges facing the government. An effective alliance among government, industry, and universities can drive China's informatization with the supply of brain power and innovative technologies while linking skills development and R&D with industrial development, so that technologies can be turned into practical and relevant applications.

In response to rapid economic and social development, the Chinese government's functions are shifting from micromanagement to macro-coordination. The big challenges will be to entrust the private sector to participate in the next wave of important e-government initiatives and to create incentives for the country's application providers to develop the capability for supporting such complex tasks. The financial incentives could include outright funding support, as well as tax holidays and interest subsidies for loans.

At the same time, the government should enforce intellectual property rights more strictly to support incentives for inventors. To encourage China-based innovations and improve R&D capacity, the country needs to tackle intellectual property protection. Like foreign companies, Chinese IT service firms, for instance, face extensive piracy challenges. Although the government has enacted tough anti-piracy laws and conducted well-publicized crackdowns, piracy rates remain high, giving firms little incentive to invest significantly in R&D to create new products. This poses a major barrier to the development of a packaged software industry to meet local demand. This is a serious concern—particularly for outsiders, who feel that there is substantial product counterfeiting and who doubt the effectiveness of enforcement mechanisms for China's software vendors. Intellectual property and copyright laws are also important to assure foreign companies that bringing new technologies and product development processes to China will not result in loss of competitive advantage.

The Chinese government recognizes the importance of improving the environment for private equity and venture investment, calling it a "critical component in promoting technology innovation" (*Red Herring* 2006). There have been calls for less-stringent listing requirements to give smaller companies—many of which are start-up high-technology enterprises—an opportunity to raise capital in equities markets.

Striking a Balance between Government Regulation and Free Market Dynamics

The last key issue is to address the complexity of regulatory scope and responsibilities in an increasingly convergent technology context. Integrated information and services are delivered across different platforms such as telephone, Internet, mobile telephony, and broadcasting. In China, eliminating further barriers to convergence will be a long process because of the entirely different histories of ownership, control, and regulation of networks.

In terms of mobile telephony, the Internet, and their manifold applications, China's regulatory approach is as complex as its institutional arrangements and can be characterized as having two seemingly unrelated tracks. On the one hand, in economic and commercial terms, China is moving toward an enabling environment conducive to ICT industry promotion and economic growth. Increasingly, this enabling environment is embracing international best practices to promote a sophisticated information society. On the other hand, China's lack of adequate privacy protection and its approach toward certain aspects of openness of information are seen as inhibitory.

This situation is changing and improving. New applications such as Short Message Service, instant messaging, and blogging have helped people form networks on a large scale and at a fast speed. Even the Party itself pays attention to the deluge of public comment. In March 2006, Prime Minister Wen Jiabao said that the government should listen "extensively" to views expressed on the Internet (*The Economist* 2006). It remains to be seen whether the market will prevail over restrictions.

There is a need to establish stable, clear, and transparent government structures to support convergence and strengthen the legal and regulatory system. The legal framework should stipulate the principles and scope of informatization, which can improve the consistency and coordination of regulations. In addition, the role of regulators and regulation itself must be reevaluated. Basic principles for regulatory

Connecting the Issues: A Summing Up

reform include encouraging market-based approaches and facilitating market entry; promoting business confidence and clarity; strengthening contract enforcement; ensuring interoperability of systems, standards, and networks; and protecting intellectual property and consumer rights.

The Chinese government needs to continue to strike the nuanced balance of government planning and regulation with the free play of market dynamics. While certain areas in the development of informatization may require some regulation, China could benefit from international experience and consider, where appropriate, either adopting industry norms or establishing default rules instead of more regulation. Self-regulation and dispute resolution can prove effective and indirectly address many of the needs identified in this publication.

In some respects the problems affecting China's ICT policies and strategies are not significantly different from those that the country will encounter in other sectors. However, the rapid pace of technology development means that ICT issues are being addressed first or more urgently than problems in other sectors. Moreover, it is clear that the sector's development will be felt throughout the economy via the adoption of technology. Thus the government's decisions about ICT can also be seen as decisions on the course of the economy as a whole.

References

BMI (Business Monitor International). 2005. "China Telecommunications Report Q4 2005." London.

CCID (China Center for Information Industry Development). 2005. "Relationship among E-government, E-commerce and E-community." Beijing.

———. 2006. "2005 China E-government Website Development Status Report." Beijing. http://www.ccidconsulting.com/2005govtop/default.shtml.

CEInet. 2004. Annual Database. Beijing. http://www1.cei.gov.cn/ce/cedb/zonghe.htm.

Chase, Michael S., Kevin L. Pollpeter, and James C. Mulvenon. 2004. "Shanghaied? The Economic and Political Implications of the Flow of Information Technology and Investment across the Taiwan Strait." RAND, http://www.rand.org/pubs/technical_reports/2005/RAND_TR133.pdf#search=%22Shanghaied%3F%20The%20Economic%20and%20Political%20Implications%20of%20the%20Flow%20of%20Information%20Technology%20and%22.

China Academy of Sciences and HP China. 2004. "Survey of Enterprise Informatization Status in China." Beijing.

China Dong Wu Ren Cai Hotline. 2005. "Ministry of Information Industry: Lack of Human Resources is an Obstacle for the IT Industry in China." Su-zhou. http://www.dwrc.sz.js.cn/dwrc/Common/winopen.asp?ID=339&Type=1.

China Ministry of Commerce. 2005. *China Foreign Investment Historical Statistics.* Beijing. http://www.mofcom.gov.cn/static/v/tongjiziliao/v.html/1.

China Mobile (Hong Kong) Limited. 2005. "Form 20-F for the Fiscal Year Ended December 31, 2004." China Mobile. http://www.chinamobileltd.com/images/pdf/CMHK_2004_20f_en.pdf.

China Netcom. 2005. http://www.china-netcom.com/english/inv/Corporate.htm.

China Newsnet. 2004. "Insight of E-government Development in China: Issues Not on 'E,' But on 'G.'" July 7. http://www.china.org.cn/chinese/zhuanti/607998.htm.

China Telecom. 2005. http://www.chinatelecom-h.com/eng/corpinfo/structure.htm.

Chinese National Commission for UNESCO (United Nations Educational, Scientific, and Cultural Organization). 2004. "Educational Development in China." UNESCO. http://www.

References

ibe.unesco.org/International/ICE47/English/Natreps/reports/china_ocr.pdf#search=%22UNESCO.%202004.%20Educational%20Development%20in%20China%22.

CNNIC (China Internet Network Information Center). 2005. "16th Statistical Survey Report on the Internet Development in China (July 2005)." CNNIC. http://www.cnnic.net.cn/download/2005/2005072601.pdf.

———. 2006. "17th Statistical Survey Report on the Internet Development in China (January 2006)." CNNIC. http://www.cnnic.net.cn/download/2006/17threport-en.pdf#search=%2217th%20Statistical%20Survey%20Report%20on%20the%20Internet%20Development%20in%20China%20%22.

David, P. 1991. "Computer and Dynamo: The Modern Productivity Paradox in a Not-Too-Distant Mirror." In *Technology and Productivity: The Challenge for Economic Policy*. Paris: Organisation for Economic Co-operation and Development.

de Jonquieres, Guy. 2006. "To Innovate, China Needs More Than Standards." *Financial Times*, July 12.

The Economist. 2006. "The Party, the People and the Power of Cyber-talk." April 27.

Farrell, Diana, Nishir Kaka, and Sascha Stürze. 2005. "Ensuring India's Off-shoring Future." *McKinsey Quarterly, 2005 Special Edition: Fulfilling India's Promise*. http://www.mckinseyquarterly.com/article_abstract_visitor.aspx?ar=1660&L2=1&L3=106.

Ge, Daokai. 2004. "The Construction and Application of Higher Education Informatization." *China Higher Education* (October).

Global Insight. 2006. "China Unicom Reports $613.9 Million Net Profit in 2005." *Global Insight Perspective*, March 24.

Guermazi, B., and D. Satola. 2005. "Creating the 'Right' Enabling Environment for ICT." In *e-Development: From Excitement to Effectiveness*, ed. Robert Schware. Washington, DC: World Bank.

Horsley, Janie P. 2004. "Shanghai Advances the Cause of Open Government Information in China." Yale Law School, China Law Center, New Haven, Connecticut. http://www.law.yale.edu/documents/pdf/Shanghai_Advances.pdf.

Huang, Ping, and Frank N. Pieke. 2003. "China Migration Country Study." U.K. Department for International Development, London.

IDC (International Data Consulting) and BSA (Business Software Alliance). 2003. "Expanding Global Economics: The Benefits of Reducing Software Piracy." BSA. http://www.bsa.org/resources/upload/IDC-White-Paper.pdf.

IDG (International Data Group). 2005. "Competition Heats Up for Dell in China's PC Market." IDG News Service, March 18. http://www.infoworld.com/article/05/03/18/HNheatfordell_1.html.

IFC (International Finance Corporation). 2005. "The ICT Landscape in the PRC: Market Trends and Investment Opportunities." Washington, DC.

ITU (International Telecommunication Union). 2006. "World Telecommunications Indicator Database." Geneva.

Letner, R. L. 2005. "China's Electronic Signatures Law Goes into Effect." http://www.dwt.com/practc/sha_chi/bulletins/04-05_ElectronicSignatures.htm.

Li, Chen. 2005. "ICT Use in Education: Policy Goals and Implementations." In Meta-survey on the Use of Technologies in Education. http://www.unescobkk.org/fileadmin/user_upload/ict/Metasurvey/CHINA.PDF.

Li, Shuhong. 2001. "The Outflow of Western Talents and Its Countermeasures." *Academic Journal of Shanxi Financial and Economic University* 23.

References

Li, Xiaohua. 2004. "Enforce Long Distance Education Project, and Promote the Leap-Frogging" Development of Rural Education." *China E-education* (August).

Lu, Zhongyuan. 2005. "The New Trend and Its New Demand for National Informatization of China's Economic and Social Development." Paper presented at the workshop, "Informatization Strategy and Economic Transformation: Trends, Experiences and Outlooks," sponsored by the World Bank and Department for International Development, Su-zhou, May 6.

McKinsey Quarterly. 2005. "Can China Compete in IT Services?" January 11. http://news.com.com/Can+China+compete+in+IT+services/2030-1069_3-5520233.html.

Microcomputer World. 2003. "The Informatization Senior Management Talents Training Project Started." November 13.

MII (Ministry of Information Industry). 2004. *Statistics Reports of Economic Running in Electronic Information Industry*. Beijing:

———. 2005. *Statistics Reports of Economic Running in Electronic Information Industry*. Beijing:

Ministry of Civil Affairs. 2000. *Outline Development Plan for IT Applications in the Civil Affairs Sector in China (2001–2005)*. Beijing:

Ministry of Commerce. 2004a. *China's E-commerce*. Beijing: Economic Science Press.

———. 2004b. *National Commodity Market System Development Report*. Beijing: Economic Science Press.

Ministry of Education. 2003. *National Statistics Yearbook of Graduate Student Enrollment, 1996–2002*. Beijing: Aviation University Press.

Ministry of Information and Communication (MIC). 2002. "Korea Vision 2006." Republic of Korea.

NBS (National Bureau of Statistics). 2003, 2004. *China Statistics Yearbook*. Beijing: China Statistics Press. http://www.stats.gov.cn/tjsj/ndsj/yb2004-c/indexch.htm.

———. 2005. *China Statistics Yearbook*. Beijing: China Statistics Press. http://www.stats.gov.cn/tjsj/ndsj/yb2004-c/indexch.htm.

———. 2003. *China Statistics Yearbook of Hi-tech Industry*. Beijing: China Statistics Press.

NBS (National Bureau of Statistics) and Ministry of Science and Technology. 2003. *China Science and Technology Statistics Yearbook*. Beijing: China Statistics Press.

OECD (Organisation for Economic Co-operation and Development). 1999. *Managing Innovation Systems*. Paris: OECD.

Park, Albert. 2004. "Rural-Urban Inequality in China." Background paper prepared for the 11th Five-Year Plan of China, World Bank, Washington, DC.

Penrose, Denise. 2005. "Education in China." Elsevier Science and Technology Books, *Electronic News*, November 7. http://www.reed-electronics.com/siliconroad/article/CA6281904.html.

People's Daily. 2005. "China's Urban-Rural Income Gap May Reach the Highest in History." December 5. http://english.people.com.cn/200512/05/eng20051205_225741.html.

Perkinson, Ron. 2005. "Beyond Secondary Education: The Promise of ICT for Higher Education and Lifelong Learning." In *E-development: From Excitement to Effectiveness*, ed. Robert Schware. Washington, DC: World Bank.

Qiang, Christine Z. 2001. "Building the Information Infrastructure." In *Seizing the 21st Century: China Knowledge Economy*, ed. C. J. Dahlman and J. Aubert, Washington, DC: World Bank.

Qiang, Christine Z., and Alexander Pitt. 2003. "Contribution of Information and Communications Technologies to Growth." Working Paper 24, World Bank, Washington, DC.

http://iris37.worldbank.org/domdoc/PRD/Othe0r/PRDDContainer.nsf/WB_ViewAttachments?ReadForm&ID=85256D2400766CC78525709E005FAB52&.

Qiang, Christine Z., George R. Clarke, and Naomi Halewood. 2006. "The Role of ICT in Doing Business." In *Information and Communications for Development 2006: Global Trends and Policies*. Washington, DC: World Bank.

Qu, Weizhi. 2005. "Opening Remarks." Speech presented at the workshop, "Informatization Strategy and Economic Transformation: Trends, Experiences and Outlooks," sponsored by the World Bank and Department for International Development, Su-zhou, May 6.

Red Herring. 2006. "China: VC is the Key." April 7. http://www.redherring.com/Article.aspx?a=16434&hed=China:%20â€˜VC%20Is%20the%20Keyâ€™.

Reed Electronics Research. 2004. *Yearbook of World Electronic Data*. Oxford, U.K.

Rong, Yi, and Jiahou Li. 2002. "The Investigation and Research of Network Information Behaviors of Middle Schools." *E-education Research* (January).

Satola, David. 2006. "Legal Aspects of Internet Governance Reform." *Information Policy* 1(2) (October). Amsterdam: IOS Press.

Satola, D., R. Sreenivasan, and L. Pavalasova. 2004, "Benchmarking Regional e-Commerce in Asia and the Pacific and Assessment of Related Regional Initiatives." In *Harmonized Development of Legal and Regulatory Systems for e-Commerce in Asia and the Pacific: Current Challenges and Capacity Building Needs*. New York: United Nations Economic and Social Commission for Asia and the Pacific.

State Information Center and China Information Association. 2003. *China Information Almanac*. Beijing: China Information Almanac Periodical Press. http://www.cia.org.cn/

Suttmeier, Richard P., Xiangkui Yao, and Alex Zixiang Tan. 2006. "Standards of Power? Technology, Institutions, and Politics in the Development of China's National Standards Strategy." The National Bureau of Asian Research. Seattle. http://nbr.org/publications/specialreport/pdf/SR10.pdf.

The Times. 2004. "China's Dark Horse to Stampede West." December 9.

TRLabs. 2006. *2005 Annual Report—Horizon*. TRLabs. http://www.trlabs.ca/trlabs/acrobat/05_annual_report.pdf.

UN (United Nations). 2005. *Global E-government Readiness Report 2005*. Division for Public Administration and Development Management. New York. http://www.unpan.org/egovernment5.asp.

UNCTAD (United Nations Conference on Trade and Development). 2006. "Foreign Direct Investment Statistics." UNCTAD. http://www.unctad.org/Templates/StartPage.asp?intItemID=2921&lang=1.

UNESCAP (United Nations Economic and Social Commission for Asia and the Pacific). 2004. "Harmonized Development of Legal and Regulatory Systems for E-commerce in Asia and the Pacific: Current Challenges and Capacity-Building Needs." Working paper, UNESCAP, New York.

UNESCO (United Nations Educational, Scientific and Cultural Organization) Institute for Statistics. 2005. http://stats.uis.unesco.org/ReportFolders/ReportFolders.aspx?CS_referer=&CS_ChosenLang=en.

WIPO (World Intellectual Property Organization). 2005. "*Patent Cooperation Treaty Statistical Indicators Report*. Geneva.

World Bank. 2004. "Policies for the 11th Five-Year Plan." Working paper, World Bank, Washington, DC.

———. 2006a. "China ICT Level and Investment Needs Assessment Survey." World Bank, Washington, D.C.

References

———. 2006b. "Imperatives of National Cyber Security." Working paper, World Bank, Washington, DC.

———. 2006c. *World Development Indicators 2006.* Washington, DC: World Bank.

Xinhua Financial Network News. 2004. "IBM Sells PC Unit to China's Lenovo for 1.75 Bln USD." December 9.

Yong, James S.L. 2005. *Enabling Public Service Innovation in the 21st Century: E-Government in Asia.* Singapore: Times Editions.

You, Wuyang, and Qin Tao. 2003. *Informatization and the Future China.* Beijing: Chinese Academy of Social Sciences.

Yu, Zhongyu. 2005. "China's IC Industry, the Status Quo and Future." China Semiconductor Industry Association. http://iis-db.stanford.edu/evnts/ 4307/Yu_Zhongyu_CSIA_slides.pdf.

Yuan, Juan. 2006. "Current Situation and Internationalization of Human Resource Flow in China." China Academy of Sciences. http://www.asiapacific.ca/analysis/pubs/pdfs/FlowChina.pdf.

Yusuf, Shahid, and Simon J. Evenett. 2002. *Can East Asia Compete? Innovation for Global Markets.* New York: Oxford University Press.

Index

Boxes, figures, notes, and tables are indicated by "b," "f," "n," and "t."

A

academic/business collaboration, 5–6, 87
Academy of Telecommunications Research, 51
acquis communautaire, 33
ADSL. *See* asymmetric digital subscriber lines
advertising, 8
agriculture, 98–99, 100*b*
 See also e-agriculture
Alibaba China, 112, 113*b*
AliPay, 113*b*
anti-piracy laws, 63
Apple Computers, 85
applications, supply-demand of, 121–22
Asia-Pacific Economic Cooperation Blueprint for Action on Electronic Commerce, 28
Association of Developers and Users of Open Source Software in Administrations and Local Communities (ADULLACT), 103
asymmetric digital subscriber lines (ADSL), 43, 44, 45*f*, 68
Australia, 104
Automotive Works Corporation, 112*b*

B

Baoshan Iron and Steel, 112*b*
blogs, 122
Board of Supervisors for Major State-Owned Enterprises, 116
brain drain, 83
broadband
 barriers to expansion of, 4
 charges, 44, 45*f*
 defined, 22*n*
 expansion barriers, 4
 growth of, 43, 44
 informatization and, 47
 infrastructure development, 3, 49–51, 120
broadcasting, 24
business/academic collaboration, 5–6
business process management, 108
business-to-business (B2B) e-commerce, 8, 16, 110–13
business-to-consumer (B2C) e-commerce, 114, 118*n*
business-to-government (B2G) e-commerce, 116
Business Transformation Enablement Program (Canada), 101

C

Canada, 101
Capital E-Shop, 115
CDMA-2000, 51, 72
China Chemical Network, 113
China Computer Association, 63
China Education and Research Network (CERNET), 84–85
China Internet Network Information Center, 114
China Langchao, 60
China Mobile, 39, 40, 41*t*, 43*f*, 51, 53*n*, 67
ChinaNet, 43
China Netcom, 39, 40, 41*t*, 42–43, 51, 68
China Railcom, 39, 40, 41*t*
China Satellite Communications Corporation, 4, 48
China Software Testing Center, 102
China Telecom, 39–40, 41*t*, 42–43, 43*f*, 48, 51, 68
China Unicom, 39, 40, 41*t*, 42, 51, 53*n*, 67
Chinese Academy of Sciences, 87
Chongqing, 99
CIOs, 81–82, 82*b*
code division multiple access (CDMA), 41, 53*n*, 70
collaboration, academic and business, 70
 See also partnerships
commerce. *See* e-commerce
competition
 international competition and cooperation, 118
 in mobile market, 41–42
 research and development and, 118

competition (*Continued*)
  in telecommunications infrastructure, 51
  value-added services, 34
computer industry
  chip export/import, 4–5
  personal computers, 4, 15, 58–60
computer viruses, 30, 64
consumer protection, 27
copyright protection, 27, 35, 122
core technologies, 5–6
CRC-Pinnacle Consulting, 64
Criminal Code, 35
customer resource management (CRM), 108, 108*b*
cyber crime, 3, 35
Cybervision Intrusion Detection and Management Systems, 65*b*

D

data networks and services, 24
Datang Mobile, 51
data protection/privacy, 3, 27, 34
demand-and-value assessment methodology, 7, 104
developing countries, ICT role in, 12
DHC, 63
dibao system, 98
digital broadcasting, 67–68
digital divide, 21–22
digital media, 5, 66*f*, 66–69
Digital Multimedia Broadcast, 74*n*
Digital Video Broadcast-Handheld, 74*n*
Directive on Government Openness, 37*n*
distance learning, 6, 79–80, 83, 84
distribution systems, 117
domestic standards, 120–21
Dongfeng Automobile Group, 112*b*
dot-coms, 110
DRAM (Dynamic Random Access Memory), 70

E

e-agriculture, 98–99, 99*b*, 100*b*, 104*n*
e-ASEAN Initiative, 28
e-commerce, 7–8, 26, 28–30, 105–18
  See also telecommunications
  applications, 105
  business-to-business e-commerce, 8, 16, 110–13
  business-to-consumer (B2C), 114, 118*n*
  changes in supplier/client contacts, 111, 111*f*
  Chinese firms and, 110–16
  drawbacks to, 114–15, 115*f*
  in East Asia and the Pacific, 28, 29*t*
  e-transactions, 27, 115–16
  improving environment of, 117
  internal informatization, 108–10
  leaders in, 112*b*
  logistics and distribution systems, 117
  manufacturing and Internet/ICT, 109*f*, 109–11, 110*f*
  obstacles, 30, 111, 113*f*
  online shopping frequency, 114, 115*f*
economic development
  economic structure, 17*f*, 17–18
  foreign direct investment, 18*f*, 18–19
  ICT and restructuring of, 12, 13
  information resource sharing, 102–3
  rural-urban population/income divide, 19*f*, 19–20
economic resources, 13
E-Contracting Convention, 30, 32
e-customs, 16, 92
education and training, 75–80
  distance education, 79–80
  ICT-supported, 84–85
  primary and secondary, 76*f*, 76–77, 77*b*, 77*f*
  universities and technical institutes, 77–78
  vocational training/certification, 78–79, 79*b*
e-fiscal management initiative, 16
e-government, 89–104
  advancing, 6–7
  development stages, 89, 90*f*
  Government Online project, 95–96, 96*b*, 96*f*
  ICT applications, 90, 91*f*
  initiatives, 2
  integrating/sharing information resources, 102–3
  IT architecture, 100–101
  local e-government and e-community, 97*f*, 97–99
  maximizing investment returns, 103–4
  promotion of, 85, 86*b*
  public-private partnerships, 101–2
  public service projects, 90–100
    Golden Projects, 92–95, 93*t*, 94*f*
  quality standards, enhancing, 102
  readiness rankings in East and South Asia, 91, 92*t*
  rural informatization case study, 99–100*b*
  transparent information flow goal, 104
e-Government Observatory, 7
encryption technologies, 30
energy industry, 20
eNet, 114
English language skills, 5
enterprise resource planning (ERP), 108, 108*b*
e-procurement, 8
E-Signature Law, 3, 15, 28, 29*b*, 29*t*, 36*b*, 117, 118*n*
e-taxation project, 16
European Union
  e-Government Observatory, 103
  ICT regulation, 33
  IT workers, 82

Index

evaluation frameworks, 7
Evolution Data Only technology, 51
export market, 119–20

F

feasibility studies, 7
Federal Enterprise Architecture (U.S.), 101
Five-Year Plan (China's 10th) Special Planning of Informatization document, 11
Five-Year Plan (China's 11th), 52
fixed-line operators, 40–41, 44, 46f, 47f, 53n
foreign direct investment, 18f, 18–19
foreign investment, 4
France, 103

G

gaming, online, 25, 68–69
General Administration of Press and Publication, 25, 69
global system for mobile communications network. *See* GSM
goals, of informatization, 20–21, 21f
Golden Agriculture, 16
Golden Gate, 16, 92–95, 93t
Golden Projects, 2, 7, 15, 92–95, 93t, 94f
Golden Shield, 92, 93t
Golden Tax, 16, 93t, 94
Golden Wealth, 16, 92, 93t
government
    See also e-government
    information access, 3, 34–35
    partnership with industry, 86
Government Online, 95–96
Great Wall Broadband, 43
Great Wall Group, 87
GSM (global system for mobile communications), 41–42, 51, 53n

H

Haier Group, 112b
hardware
    integrated circuit industry, 56b, 56–58
    personal computer industry, 58–60
High Speed Downlink Packet Access, 53n
Hong Kong, 18
Huawei, 73
human capital, 5
human resources, 75–88
    challenges to, 80–83, 81f
        brain drain, 83
        lack of ICT management expertise, 80–82
        lack of public awareness/application capacity, 80
        supply-and-demand gap, 82–83
    ICT education and training, 75–80
        distance education, 79–80, 84–85
        primary and secondary education, 76f, 76–77, 77b, 77f
        universities and technical institutions, 77–78
        vocational training/certification programs, 78–79, 79b, 80b, 82b
    improving, 6
    information management, 87–88
    local capacity building and, 14
    public awareness, raising, 85–86

I

IBM, 60b, 79b, 86
ICT (information and communications technology)
    characteristics of, 12–13b
    defined, 1
    development of, 4–6, 11, 13
    digital broadcasting, 67–68
    digital media, 66–69
    e-business and, 7–8
    as economic value source, 12
    e-government and, 6–7
    hardware, 56–60
    human resources. *See* human resources
    information and network security services, 63–66
    infrastructure, 11, 13
        See also telecommunications
    innovation and, 4–6, 69–74
    mainstreaming, in education/training systems, 86–87
    mobile data services, 67
    online games, 68–69
    personal computer industry, 58–60
    product export potential, 119
    professionals, 82–83, 83t
    public awareness of, 85–86, 86b
    regulatory agencies. *See* regulatory environment
    role of, in development strategies, 2
    rural access to, 48
    skills needed, 22
    social effect of, 1
    software, 60–63
    standard development/international practice alignment, 72–73
    worker shortage, 6
immigration, 19
income ratio, 19f, 19–20
incubator companies, government funding of, 4, 8
industrialization, 19
information retrieval, abilities of students, 77b
information and communications technology. *See* ICT
information management, 87–88
information network security, 63–66, 64f, 65b

information resources, 7, 102–3
Information Technology Security Certification Center, 66
informatization
   China's framework for, 20–22, 21f
   defined, 11
   e-community, 97f, 97–98
   economic development and, 6, 17–20
      economic structure, 17–18
      foreign direct investment, 18f, 18–19
      urbanization and rural-urban income divide, 19f, 19–20
   e-government initiatives, 116–17
   enablers and building blocks, 2–6
   enterprise level (e-business), 105
   at enterprises, 106, 106f
   goals, 106
   human resource levels critical to, 6
   ICT development stages, 13
   impediments to, 21–22
   industrialization and, 15
   institutional reform and, 9
   internal applications of, 105
   investments, by industry, 107, 107f
   objectives, 20, 106, 106f
   obstacles, 119
   regulatory issues, 27, 27t, 118n
      See also regulatory environment
   since mid 1990s, 15–16
   strategy, 2, 11–22, 14f
      framework, 20–22, 21f
      global picture, 11–15
innovation
   barriers to, 121
   for domestic/foreign markets, 119–20
   government role in, 120
   ICT and, 5, 12
   strategies, 69–74
   technology innovations, 12
instant messaging, 122
Institute of Electrical and Electronic Engineers, 72
integrated circuit industry, 4–5, 56b, 56–58, 57b, 57f
intellectual property rights
   piracy concerns, 5, 35, 63, 121
   regulatory obstacles and, 27, 69
internal informatization, 108b, 108–10
international production networks, 19
International Telecommunication Union (ITU), 72
Internet
   broadband infrastructure development, 4, 49–51
   China's use of, 3
   content development/regulation, 3, 34, 44
   development/regulation, 24, 28–30
   fees as percent of income, 44, 45f
   Internet service providers (ISPs), 43
   manufacturing industry and, 109, 109f
   operators, 43
   penetration in urban/rural areas, 47, 47f
   resolution on Internet security, 30
   usage, 22, 80, 81f
Internet Engineering Task Force, 72
Internet Protocol television, 68
Internet Society image coding standards, 72
internships, 87
Ireland, 13, 103
IT industry, 15, 16

J

Japan, 74n
Jiangming, 65b
JinXiaoCun (JXC), 108
Jitong Corporation, 39
joint ventures, 19
joyo.com, 114

K

Ken Si Jie, 102
Kingdee, 63
Kingsoft, 65b, 69
Korea, Republic of, 13, 70

L

labor competitiveness, 119–20
leapfrogging, 12
legal environment
   See also regulatory environment
   anti-piracy laws, 63, 121–22
   e-signature law, 29b
   imbalance of rights/obligations, 3
   informatization issues, 27, 27t, 122–23
   legislative mechanisms, 36b
   recent laws/regulations, 28–30, 31t
Lenovo, 60b, 73, 85, 87, 112b
literacy, 75
Liu Chuanzhi, 60b
Local Authority Software Consortium (UK), 7
Local Government Computer Services Board (Ireland), 7
logistics systems, 117
low-frequency mobile technology, 50

M

management information systems, 108
manufacturing industries, 12, 20, 109, 109f
marketing, 8
materials requirement planning (MRP), 108
m-commerce, 116
MediaFLO, 74n

Index

Mexico, 86
Microsoft, 86
migrant workers, 97
MII. *See* Ministry of Information Industry
Ministry of Culture, 25
Ministry of Information Industry (MII), 3–4, 24, 48, 51, 52, 67, 71
Ministry of Public Security, 63, 66
Ministry of Science and Technology, 71
minority groups, 22
mobile data, 67, 74n
mobile phones
  market shares, 43f, 44
  mobile data services, 67
  numbers of, 3
  operators, 41–43, 53n
  regional use of, 44, 46f, 47
  transactions using, 116
mobile virtual network operators, 51
modernization, 13
monitoring frameworks, 7
Monternet, 67
Multimedia Messaging Service (MMS), 66
Multi-Service Ring, 72
My8488 E-commerce Co., 114

N

National Computer Network Emergency Response Technical Team, 65
National Computer Rank Examination, 78
National Development and Reform Commission, 25
National Economic Information Joint Committee, 15
National People's Congress, 29b
Netease, 69
network security, 3, 27, 34, 63–66, 64f, 65b

O

office automation, 108, 108b
Office of Communications (UK), 33
Olympics of 2008, 68
online gaming market, 68–69
online transactions, 8, 16, 28, 114–16, 115f, 117
operating systems, 78
Oracle, 86
Outline of the National Informatization Development Plan (1997), 11, 15
outsourcing industry, 5, 62–63

P

partnerships
  government-industry, 86
  ICT success and, 14
  public-private, 86, 101–2
patents, 1, 12, 71

Performance Reference Model (U.S.), 104
personal computers (PCs), 58–60
  incentives to produce inexpensive, 119
  market shares, 59, 59f
  ownership of, 4, 49
  penetration by province (2003), 49, 50f
  top global producers, 58, 59f
personal handy phone system (PHS), 42–43, 53n
PetroChina, 112b
piracy, 5, 35, 63, 121
population, ICT distribution, 22
privacy protection, 23, 27, 34
product focus, 62
production chains, 12
productivity, 12
Program 863, 116
public awareness, of ICT, 85–86, 86b
public-private partnerships, 86, 101–2, 105
public service projects, 7, 90–100

Q

Qing Hua Wang Bo, 102
QUALCOMM, 51, 53n, 74n

R

real estate, 17
reform, 32–35
regional dispersion, 62
regulatory environment, 23–37, 25t
  areas addressed by, 28–30
  balancing rights/obligations, 36–37
  challenges to, 30, 32
  digital media and, 5
  in EU and UK, 33, 33b
  free market dynamics and, 122–23
  ICT deployment and, 14
  ICT regulation/regulatory institutions, 23–25, 26f
  improving certainty/coordination, 35–36
  informatization issues, 27, 27t
  laws and regulations, 30, 31t
  network convergence and, 4, 52–53
  new technology obstacles, 69–70
  proposed framework for, 25–27
  reform, 26, 32–35
  resolution on Internet security, 30
research and development
  competitiveness and, 118
  focus on relevant applications, 5–6
  foreign direct investment and, 18
  informatization focus, 120
  intellectual property and copyright laws, 27
  piracy concerns, 121–22
  spending and ICT patent applications, 71, 72f
  stimulating investment in, 71–72

135

research and development (*Continued*)
  strategic technologies and, 70
  support needed for, 5
  TRLabs, 70, 71*b*
Rising Technology, 65*b*
rural areas
  development of, 16
  online information and, 7, 98
  providing universal access to, 48–49
rural to urban economy, 101

S

secrecy, culture of, 7, 104, 104*n*
security, information and network, 27, 63–66
service industries and economies, 12, 13, 17, 20
Shanda, 69
Shanghai General Motors, 112*b*
Shanghai Shenxin Information Technology Academy, 79
Short Message Service, 66, 122
Sina, 114
small and medium-size enterprises (SMEs), 8, 65, 111, 114, 117–18
SMS, 67
smuggling, 92
social development, 13
social services
  disadvantaged groups and, 98
  software investment, 107
software, open source, 35
software industry, 60–63
  China's software market growth, 61, 62*f*
  Chinese software parks, 61, 61*t*
  features of, 5, 61–63
  mobile data software, 74*n*
  social services and, 107
spam, 64
standards, domestic, 72–73, 120–21
Standing Commission, 36*b*
startup companies, government funding of, 4
State Administration of Radio, Film and Television, 24, 52, 68
State Council Informatization Leading Group, 23, 81
State Council Informatization Office, 95
stock exchanges, 28
synchronous code division multiple access (SCDMA) technology, 50

T

Taiwan, 18
taxation, of online transactions, 117
TDX telephone switches, 70
technology
  domestic standards and, 120–21
  effective implementation of, 14
  foreign firms/market access and, 4
  improving production/demand links, 73–74
  innovations, 12
  reliance on foreign, 22
techno-nationalism, 120
telecommunications
  employees/revenues of main providers, 40, 42*f*
  infrastructure, 3–4, 15, 39–53
    broadband, 49–51
    fixed-line operators, 40–41
    Internet operators, 43
    mobile phone operators, 41–43
    sector performance, 44–47
    sharing, 51
    strategic focus, 47–53
    3G networks, 51
    universal access, 48–49
  institutional structure of, 24, 24*f*
  investment promotion, 52
  investment/revenue (1990–2004), 39, 40*f*
  laws/regulations, 24, 28
  main providers, by market segment, 39, 41*t*
  network convergence, 52–53
  penetration in China (1994–2004), 44, 44*f*
  penetration in developing economies (2004), 44, 46*t*
  reform, 32–35
  WTO commitments, 40, 42*t*
Telecommunications Regulations, 28
telephone service, 3–4
  See also mobile phones
television broadcasting, 67–68
3G telecommunications networks, 47, 51, 53*n*, 72
time division-synchronous code division multiple access (TD-SCDMA), 51, 53*n*, 72
TopSec, 65*b*
TQDigital, 69
training. *See* education and training
TRLabs, 70, 71*b*
Tsinghua University, 81

U

UFSoft, 62
Uni-Info, 67
Union of National Teachers Education Network, 80*b*
United Kingdom, 33, 33*b*, 103
United Nations Commission on International Trade Law, 30
United States
  e-government applications, 101, 104
  IT-related jobs, 82
  software exports, 74*n*
universal access, 48–49

Index

universities. *See* education and training
university-business links, 87
urbanization, 19–20, 19*f*

V

value added taxes, 16
very small aperture terminals (VSAT), 49
vocational training, 78–79, 79*b*
Voice over Internet Protocol (VoIP), 22*n*, 40

W

WAPI, 73
Web sites
    e-commerce, 107, 110–11
    government, 94–96, 95*t*, 96*b*, 96*f*, 104*n*
Wen Jiabao, 122
wideband-code-division multiple access (W-CDMA), 51, 72
wireless data services, 67
Wireless Local Area Network (WLAN), 47, 73
World Trade Organization, 7, 105

Y

Yahoo! China, 113*b*

Z

ZTE, 51, 73

DATE DUE

The World ... sts and natural resourc... rint ***China's Informatio***... st-consumer waste, in ac... r usage set by the Green ... ublishers in using fiber ... his paper, the following w... 1 gallons of wastewater, ... ion BTUs of total energy ... ve.org.

Demco, Inc. 38-293